Gender and Diversity in a Problem and Project Based Learning Environment

Gender and Diversity in a Problem and Project Based Learning Environment

Xiang-Yun Du

Aalborg University
Denmark

LONDON AND NEW YORK

Published 2012 by River Publishers
River Publishers
Alsbjergvej 10, 9260 Gistrup, Denmark
www.riverpublishers.com

Distributed exclusively by Routledge
4 Park Square, Milton Park, Abingdon, Oxon OX14 4RN
605 Third Avenue, New York, NY 10158

First published in paperback 2024

Gender and Diversity in a Problem and Project Based Learning Environment / by Xiang-Yun Du.

Routledge is an imprint of the Taylor & Francis Group, an informa business

ISBN: 978-87-92329-84-4 (hbk)
ISBN: 978-87-7004-525-4 (pbk)
ISBN: 978-1-003-33826-0 (ebk)

DOI: 10.1201/9781003338260

Contents

1

Introduction

An historical review (Davis, 1996) of engineering practice summarises three shared characteristic elements: (1) mathematics and natural science are central to engineering practice; (2) physical objects are the principal concern instead of people; (3) engineering does not seek to understand the world but to remake it. Accordingly, engineers are people who possess education, knowledge and skills in an engineering specialty that exceeds those of the general public and professionals who practice engineering. Engineers are very often described as factual, logical, rational, mathematical and problem-solving people (Wright, 1994).

However, with the enormous changes in technology, society and daily life, fundamental changes have taken place in engineering practice. In the globalised and network-based knowledge society, engineering skills and competencies are facing challenges when problems are becoming increasingly ill-defined and complex with the involvement of a growth of various issues such as economy, culture, sustainability and society. Added to the traditionally technical-solution-focused professional competencies, there is a growth in the demand for analytical problem-solving, knowledge sharing, collaboration, creativity and interdisciplinarity in the engineering field, which encourages an increasing variety of engineering knowledge and skills. At the same time, in spite of the great need for engineers and engineering strengths, a general lack of interest in studying engineering and science among young people has been observed in many western countries in Europe, Australia and North America.

Gender and Diversity in a Problem and Project Based Learning Environment, 1–8.

Accordingly, there is the broad recognition that engineering education needs to change in order to meet the challenges in knowledge society (Felder, 2006). As stated in the recent publication on 'Educating Engineers' (Sheppard et al., 2009), the most central change to engineering practice is a change from a linear conception of problem analysis and problem-solving that presupposed a more stable organisational and physical environments to a network, web, or system understanding of engineering work (p. 4). This change also calls for the need for graduate engineers to gain new engineering knowledge that is associated with society and humanity, and to have the competencies of communication, collaboration, innovation and professional readiness.

In 'The Engineer of 2020' (NAE, 2004), the National Academy of Engineering (NAE), USA describes attributes that are built on strengths inherited from the past while incorporating the qualities that are becoming critical in the changing world of engineering practice. 'Strong analytical skills', 'practical ingenuity', 'creativity', 'communication', 'mastery of principles of business and management', 'leadership', 'professionalism', 'high ethical standards', 'dynamics, agility, resilience, and flexibility' and 'life-long learners'. These attributes have also been included in the accreditation of engineering programs, for example, The American ABET criteria (2007), The ERU-ACE framework in Europe (ENAEE, 2005), Engineers Australia policy on accreditation (Engineers Australia, 2005). Therefore the tasks of engineering education are not only to identify specific engineering knowledge, skills and values that students need as they enter the profession, but also to determine what kind of educational experience and what approximations of professional practice will best position students to continue to develop (Sheppard et al. 2009, p. 10).

In addition to the increasing demand of diversity in engineering competences, it is time for engineering education to change and attract a diversity of students in order to take full advantage of the talents in society and to include a variety of values and contributions for the innovation of engineering work. In response to this, a growth of effort needs to be made to increase participation of students from diverse backgrounds. This also includes women in the engineering field (The Danish Ministry of Education, 2005).

The concern with regards to diversity in engineering was initiated from the feminist works and has been extended to a broad sense of participation regardless of ethnicity, race, class, religion, etc. The feminism concerns include not only the social equity but also women's potential contribution to the creativity and diversity of the science and technology work (Fox et al., 2009). Historically engineering has been a homogeneous group dominated by a certain group of men (Sagebiel and Dahmen, 2006; Male et al., 2009). Previous feminist studies have identified the following historical reasons leading to women's under represented participation in engineering: (1) women's inappropriate gender role, which keeps the ideology of femininity distant from technology and engineering (Gherardi, 1995; Kvande, 1999; Powell et al., 2008; Sulaiman and AlMuftah, 2010), (2) gender stereotypes in labour division, which defines engineering as a male-oriented profession (Kvande, 1999; Hodgkinson and Hamill, 2006), and (3) the traditional lecturer based learning environment at engineering programs overweighs sophisticated natural science knowledge and hard-core technological skills, which mainly favours male interest and expectations and ignore women's experiences and concerns (McIlwee and Robinson, 1992; Anderson and Gilbride, 2007; Ihsen and Gebauer, 2009). To different extents, these historical factors shape young people's educational choices and result in gender disproportionality — men's excessive participation in natural sciences, mathematics, and engineering and women's high participation in arts, humanities and social sciences (Smyth, 2005). Despite their academic qualifications and performance, girls and women markedly do not continue to study mathematics and natural science, and even lower proportions of women study engineering (Seymour and Hewitt, 1997; Franchetti et al., 2010).

From a feminist point of view, this historical imbalanced participation in engineering has led to unequal opportunities and uneven power distributions (Gherardi, 1995; Kvande, 1999; Sagebiel and Dahmen, 2006; Ihsen and Gebauer, 2009). Recent feminist works also argue that by inviting more female participation in engineering with their values and contributions, the engineering profession can increase its ability to meet the increased diversity of demands across society (e.g., diversity of skills, variety of products and diverse groups of customers) (Ihsen, 2005).

How to increase the diversity of engineering education? The past decades have witnessed substantial efforts to increase women's participation through recruitment activities. Most of these activities are carried out through policy development and attempts at curriculum reform. Examples can be found in Sweden (Brandell, 1996; Salminen-Karlsson, 2002), the UK (William, 2002), Portugal (William, 2002), the Netherlands (Hermanussen and Booy, 2002),

Australia (Miliszewska and Moore, 2010) and USA (Tonso, 1996; Shull and Weiner, 2002; Kim et al., 2011), etc. Various initiatives have been put in place — role-modelling, 'hands-on'interaction with science in the classroom, single-sex programs, teachers training in gender issues, etc. Recent initiatives also witnessed efforts of change of curriculum, for example, providing specific diversity lectures to all students (Schaefer, 2006); intergrading diversity concept into all engineering programs (Ihsen and Gebauer, 2009); increasing the interdisciplinarity of engineering programs by introducing humanistic elements to technology (Thaler and Zorn, 2010), implementing student-centred environment and contextualized learning context, for example, by using Problem and Project Based Learning form (Du, 2006a,c; Du and Kolmos, 2009). The expectation that these would increase the appeal of higher technical studies for girls has partly been met. However, these educational reforms alone have not been the key to a larger intake, retention and output of girls (Hermanussen and Booy, 2002). Many institutions have experienced that the measures taken until then had been insufficient to bridge the large gap between girls and technology. As it is suggested, increasing the percentage of women in enrolment does not automatically solve the problems of the male hegemony of the education (Salminen–Karlsson, 2002); to really make engineering education female-friendly, a bigger social change is needed — all the elements like curriculum content, teaching methods and the prevailing culture in engineering should be changed (Salminen–Karlsson, 2002; Daudt and Salgado, 2005; Du, 2006b; Du and Kolmos, 2009).

What constitutes a female-friendly curriculum and learning environment so as to increase the gender diversity of engineering education? It has been suggested that women's way of knowing and learning are very often characterised by the vocabulary of 'connection (with other people and the society)' and 'collective and communicative ways' (Hayes, 2000). Accordingly, some feminist scholars argue that women's learning is better supported in an environment that is different from those in traditional higher education (Hayes, 2000; Du and Kolmos, 2009). Based on gender studies in engineering as well as research in relation to new skills for engineers, further proposals can be made regarding an attractive curriculum for female students:

- Replacing a certain amount of technical content with opportunities for the application and development of practice skills, interdisciplinary knowledge and process competencies such as

management, personal and interpersonal communication and collaboration (Daudt and Salgado, 2005; Ihsen, 2005; Ihsen and Gebauer, 2009; Du and Kolmos, 2009). These are also skills favoured by employers in industries (Mosby, 2005). The most efficient way of developing these skills are through innovative pedagogy such as problem-based learning instead of traditionally used, lecture-centred top-down approach (Felder, 2006).
- Changing the traditional written examination-forms-based assessment methods, which mainly focused on technical contents knowledge and technical skills. Research suggests that assessment measures which require a broader range of skills identified by employers in practice match the preferences by female students: the interest in social implications, the concern for human needs and environmental issues, the consideration of context and relationships and the preference for collaborative learning (Du, 2006a; Mosby, 2005). Equal attention should be paid to the importance of self-evaluation on appreciation and satisfaction of learning (Du, 2006a; Du and Kolmos, 2009).
- Development of life-long learning skills so that students are equipped with the ability of learning to learn and get prepared for the changing demands of professional practice (Du, 2006a; Sheppard et al., 2009).

One major assumption from feminist works was that a friendly learning environment can be an option to encourage the recruitment of more women engineering students. How can it be provided in engineering education in order to meet the demands of diversity? As an example of a student-centred learning environment as well as contextualised approaches to instruction, Problem Based and Project Based Learning (PBL) is progressively recognised as a useful concept internationally (Kolmos and de Graaff, 2007). In particular, PBL has provided solutions to the tasks of engineering education both philosophically and pedagogically. In an international context, the effectiveness of PBL has been documented from two perspectives: students' learning and institutional development. Firstly, positive effects of PBL on students learning have been identified in the following aspects: (1) promoting deep approaches of learning instead of surface approach (Biggs, 2003), (2) improving active learning (de Graaff and Cowdroy, 1997; Du, 2006a), (3) developing the criticality of learners (de Graaff and Cowdroy, 1997; Savin-Baden, 2000), (4) improving self-directed learning capability (Hmelo and Lin, 2000; Du, 2006a), (5) increasing the consideration of interdisciplinary knowledge and skills (Kjaersdam, 1994), (6) developing management, collaboration, and communication skills (Mosby, 2005; Du, 2006b), (7) developing professional identity and responsibility development (Schmidt, 1983; Hmelo and Lin, 2000; Du, 2006c), (8) improving the meaningfulness of learning (Du, 2006a). Secondly, at the institutional level, the shift to PBL will benefit the university/departments (Kolmos and de Graaff, 2007) in terms of (1) decreasing drop-out rates and increasing rate of on-time completion of study, (2) supporting development of new competencies for both teaching staff and students, (3) promoting a motivating and friendly learning environment and (4) accentuating institutional profile.

With this background, questions can be raised whether the increased recruitment of women and can been as an improvement to gender diversity, and whether and in which ways a student-centred learning environment (PBL) can be used as an educational strategy for the increase of gender diversity in engineering education in a Danish context.

This book is based on a research on gendered learning experiences in engineering education in a PBL environment. This book reports how engineering students experience their learning processes in a student-centred learning environment and how these processes are influenced by gender relations. It documents the learning processes of both male and female students in two different engineering programs in Aalborg University, where a problem-based and project-organized learning environment is provided. Specifically, this study gives weights to learners' perceptions on learning — how they seek the meanings of learning and how they manage their identity work through learning processes. Another focus in this study is on gender, which derives from the assumption that gender relations play an influential role in the practice and experiences of learners in the context of engineering education. This book also reflects on whether and in which ways a PBL environment can benefit the increase of gender and diversity in engineering education.

The rest of the book is structured in the following way:

Chapter 2 provides a brief review of engineering and engineering education in western countries in general and more specifically in Denmark. Followed by it is an introduction to the problem-based and project-organized learning environment as educational theories and models as well as its practices in engineering education at Aalborg University. Also presented and discussed is a review of the research that has been done on engineering education in a gender perspective. The general aim of this chapter is to make the cultural context of this research understandable.

Chapter 3 starts with a review of different theoretical perspectives on learning. With a specific concern on seeking meanings and identity development in learning, this study takes its theoretical departure from learning theories that are derived from a constructivism-sociocultural approach. An analytical tool is developed to understand learning from three levels: personal level, community/organizational level, and social/cultural level.

Chapter 4 presents and discusses the basic concepts of gender and gender relations as well as their relations to knowledge and learning. The elaboration of the concept of gender starts from a brief review of different approaches that have been developed in concern with research on gender. The 'doing gender' concept is employed as a point of departure to understand gender in the social context. This is followed by a discussion on the social construction of gender relations from three levels: cultural level, institutional level, and individual level. Also introduced in this chapter is the theory of 'gender knowledge system' that relates the concepts of gender, gender relations to knowledge and learning.

Chapter 5 relates learning theories from the constructivist-sociocultural approach to gender theories from a post-structural perspective. Based on this, an analytical framework is established and elaborated through discussions of the integration of the three areas of learning, gender and engineering education. Power and change as two essential concepts that are embedded in the individual-community-society relationships are discussed from the perspectives of both learning and gender relations.

Chapter 6 reports the process of conducting this research, from initiating the study, making decisions on where to locate this research in paradigms, choosing research topics, doing explorative study, constructing of theoretical framework, designing research methods, data generation and analysis process as well as writing. Also documented in this chapter is the researcher's reflection of the learning journey as well as an analysis of the limitations and weaknesses of this research.

Chapter 7 and 8 report the empirical work in the two research sites, Electrical & Electronics Engineering and Architecture & Design Engineering respectively. In both of the chapters, the depiction starts with a brief introduction to the study programs, student's choice of engineering and the specific program, their experiences of being newcomers, their perceptions of learning engineering, and the learning processes of studying engineering in the PBL environment along the years. Also described is the learning culture of the study programs in a gender perspective regarding gendered proportion, gendered features in the learning process, and the process of developing professional identity. A gender perspective in the research process and also questions to students regarding their perceptions on gender relations lead to the analysis of gendered experiences of learning engineering in a PBL environment.

In Chapter 9, empirical findings from the two research sites are summarized, compared to each other before they are interpreted in relation to the theoretical frameworks. The analysis process is also a way to respond to the research questions.

This book winds up by Chapter 10 with conclusions and reflections on gender inclusiveness in engineering education, and on-going changes facing the development of engineering education.

2

PBL Environment for Gender Diversity in Engineering Education

The chapter aims to provide a general introduction to the research context in my study: PBL environment in engineering education from a gender perspective. It starts with a brief description of the changing process in engineering education. This transformation brings about a growing need for increasing diversity (for example diverse competences and diverse social groups). In order to prepare students with diverse engineering competences, engineering educational institutions are in an undergoing process of transition from a traditional teaching-centered paradigm to a new learning-centered paradigm. Therefore, problem-based learning as an example of contextualized learning environment that has been well-applied in Denmark is chosen as a practical research context to examine engineering education in my study. In addition, I also argue that gender, as one aspect of the diversity issues, has seldom been addressed in the forum of engineering education in Denmark. With a brief review of feminist arguments and analyses on gender issues in engineering education in different western countries, this chapter gives some examples of masculinity as the domain of the education. Some identified reasons that led to women's historical under representation in engineering education will be addressed. This chapter ends with a discussion of criteria for a friendly learning environment in a gender perspective, which establishes a platform for the analysis of my research.

Gender and Diversity in a Problem and Project Based Learning Environment, 9–30.

2.1 Engineering Education as a Research Context

This section provides a brief introduction to the undergoing changes in the development of engineering and engineering education. Some exemplified practice in the Danish context indicates a general trend for increasing diversity in engineering education.

2.1.1 A Changing Process in Engineering

Engineering has evolved and developed as a practical art and a profession over more than 5000 years of recorded history in western countries (Wright, 1994). Ways of defining engineering vary in different countries. For example, according to the Accreditation Board for Engineering and Technology (ABET), U.S. engineering is defined as 'a profession in which a knowledge of the mathematical and natural sciences gained by study, experience, and practice is applied with judgment to develop ways to utilize, economically, the materials and forces of nature for the benefit of mankind'(Wright, 1994:46). In U.K. engineering is officially regarded as 'the practice of creating and sustaining services, systems, devices, machines, structures, processes and products to improve the quality of life, getting things done effectively and efficiently' (Engineering Council, 2000:9).

Two distinctive components can be seen from different versions of engineering: engineering knowledge (which involves the 'know what') and engineering process (which involves the 'know how'). This means that the engineer's knowledge comes not only from study, but also from experience and practice (Wright, 1994). Therefore, professional engineers are expected to possess education, knowledge, and skills in an engineering specialty that exceeds those of the general public.

Engineers are very often described as factual, logical, rational, mathematical, and problem-solving people (Wright, 1994). However, in many western cultures (for example, in Denmark), engineers also have an image of having a narrow technical focus and showing no interest in society. A portray of the traditional engineer, as described in Beder's (1998) article, is 'a nerdy-looking character, with thick glasses, short hair, several pens and pencils in his shirt pocket, perhaps in a plastic pocket protector, wearing clothes that are never quite up to fashion'.

Under the impact of globalization, the roles of engineers and what kind of technology they produce and innovate are being reshaped. Accordingly, the traditional image of engineers that was associated only with mathematics and physics theories and technical tools is facing challenges. New types of engineers are called for in the new era in different western countries (Hadi, 2001; Fromentin and Werra, 2001; Lucena and Downey, 2001; Antonini, 2002; Hernaut, 2002; Barsony, 2002; Kjaersdam, 2002; Felder, 2006; Ihsen and Gebauer, 2009): besides the specialized qualifications with regard to science and technology, engineers need to be creative, innovative, flexible, eager to learn, and capable of project management and cooperative in multidisciplinary teamwork. They need operate with market and customers in mind, satisfy society's needs and conserve resources and nature, and take responsibility for the application of technology supplied by them. As an integral part of a global society, future engineers are expected to master a combination of diverse capabilities and be aware of the importance of diversity. These new engineering competences are also expected in the Danish context (Andersen and Hansen, 2001; Vidal, 2001; IDA, 2002; Kjarsdam, 2002; Kolmos, 2002a; Du, 2006c; Du and Kolmos, 2009).

2.1.2 Transformations in Engineering Education

In the current western societies, higher education (universities) is facing challenges for deep transformation in the changing process of knowledge creation and dissemination. Used to be the key institution in the production and reproduction of formalized knowledge and high level expertise for the modern society (Barnett, 1994), universities, as learning organizations, are undergoing diverse changes in order to provide sufficient chances for the members to learn. Among all the changes, the shift of the core of education from teaching to learning stands out as one of the significances (Barnett, 1994; Bowden and Marton, 1998; Jarvis, 1992, 2001; Kolmos, 2002). Researchers (Barnett, 1994; Bowden and Marton, 1998) on higher education have challenged the concepts of 'skills', 'quality' 'capability', 'expertise' that used to orientate modern universities and proposed new vocabularies like 'understanding', 'meanings', and 'transformation'. The importance of personal growth is also suggested to be brought into the concern of the development of universities (Jarvis, 1992, 2001, 2009; Illeris, 2007, 2009).

In relation to engineering education, this trend, combined with the changing processes in engineering profession brings about need for engineering education to transform in order to facilitate supportive learning milieu and provide qualified graduates. Responding to this, Kolmos (1992, 1996) proposes 'open learning' concept in engineering education. Four aspects are suggested for the reform in educational practices at engineering universities, which are respectively learner-centered, social, problem-oriented and interdisciplinarity, projects as methods. According to Kolmos (1996), the introduction of 'open learning' into the theoretical thinking in educational education are to meet the following needs: pedagogical regard to new target groups, new qualification requirements from industry and new methods of measuring educational quality. Therefore, in order to facilitate 'open learning', educational change is a challenge and a trend facing engineering education in western countries in general.

Kolmos (2002a) summarizes two general methods in knowledge creation and transition in engineering education in western countries: traditional approach and techno-science approach. The traditional progression is mostly occupied with basic science, natural science, and technology-related courses, and consists of limited application-related content. Based on this, universities and companies take fundamentally different responsibilities in the process of technological knowledge production and organization. Universities are responsible for educating engineers with knowledge of the basic principles in technical science, whereas companies are responsible for transferring this knowledge to technological innovations.

The techno-science approach in engineering education is based on contextual learning, and is focused on innovation and related to companies and markets. In this approach, the concept of knowledge is broader (covering knowing why, what, how and who), interdisciplinary (incorporating new combination of existing knowledge and in relation to new fields of application), and contextual (including social dimensions). Accordingly, the production of knowledge is flexible and dynamic.

Both of these two approaches are undergoing changes.[1] In the techno-science approach, engineering educational programs have started to develop

[1] Kolmos (2002a) discusses undergoing changes in the two approaches respectively, however, in my research, I only focus on student-centred learning environment in the techno-science approach.

more student-centred curricula, where learners are expected to develop engineering competences[2] that combine both contents and methodical competencies (technical competencies, interdisciplinary perspective, problem analysis and problem solving) and process and organizational competencies (communication, project management, teamwork, self-management, organization). In practice, different educational models are developed to implement this transformation, an example of which, the problem-based and project-organized model will be discussed in the next section.

2.1.3 Present Practices in Denmark

Ways of defining engineering education and engineers vary in different countries. In my study, I will mainly use these concepts based on their cultural meanings in the Danish context. In Denmark, according to the principles of the Danish engineer society (IDA) the official title of engineers is referred to those who have certain levels of engineering education (Bachelor and Master level). Accordingly, in my study, 'engineering education' refers to the Danish education that turn out two levels of engineers: 'diplomingenior' with a bachelor degree based on three-and-half-year education, and 'civilingenior' with a master degree based on five-year education. There are altogether eight educational institutions providing engineering education. Among them, three are at university level, offering both bachelor and master degree education, and five colleges, offering bachelor degree education. At all these engineering educational institutions, there are about 17,000 engineering students and approximately 2550 members of staff (Kolmos and Vinther, 2004).

In the past years, some actions have been taken in different engineering educational institutions in Denmark in the transformation process. Firstly, new study forms have been developed to establish student-centered and workplace-imitated learning environments. Secondly, new programs[3] have

[2] The terminology of 'competence' and 'skill' can be confusing. In my study, the term 'competence' is used in the setting of education, and according to Kolmos (2002), competences involve some other capabilities, in addition to the concern of knowledge and skills, such as the ability of transferring knowledge and skills to practice, capacity of learning in different situations, capability of solving problem in interdisciplinary approaches, as well as cooperation and communication abilities, etc. In this sense, 'competence' has a broader meaning than 'skill'.

[3] For example, Health Technology and Architecture & Design at Aalborg University (AAU), and Food Science Engineering at Technical University of Denmark (DTU), which are designed for the need of the society and for recruitment of young people in general.

been introduced to enrich the contents of engineering disciplines. Thirdly, new expectations have been set up to broaden the engineering skills and competences, which are indicated in the principles of the framework provisions for the engineering study programs (see appendix 4). Indicators are the new vocabularies like 'creative problem-solving', 'communication', 'co-operation and group work', 'management' 'international work environment', etc. in addition to 'understanding of science theories', 'application of modern information technology', 'logical reasoning', 'tackle new problems'. A general trend of increasing diversity in the development of engineering education in Denmark can be witnessed from these undergoing changes (Kolmos, 2006).

2.2 PBL Environment as a Research Context

In recent years, most of the engineering educational institutions in Denmark are in an undergoing process of transformation from traditional paradigm, which is discipline-oriented, basic and applied technical knowledge-based, and lecture-centered, to a new paradigm, which is more interdisciplinary, contextualized, a complex understanding of technological knowledge-based, and student-centered, through implementing a problem-based and project-organized curriculum (Kolmos, 2002a; Graaff and Kolmos, 2003; Kolmos and Vinther, 2004; Mosby, 2005). This provides a starting point for the choice of PBL environment as the research context in my study. This section will provide a brief instruction to the PBL concept regarding its theoretical background and practice in the context of Aalborg University as an educational model.

2.2.1 Problem-based Learning as an Example of Student-centred Learning

As an example of student-centered learning environment as well as contextualized approaches to instruction, problem-based Learning[4] (PBL) is more and more accepted as a useful concept in education (Graaff and Cowdroy, 1997; Hmelo and Evensen, 2000; Kolmos and Vinther, 2004; Mosby, 2005; Du, 2006c; Du et al., 2009). This is because the most important innovative aspect of this educational concept is 'the shift from teaching to learning, and

[4]Problem-solving as a concept in curriculum design had its original inspiration from works of Dewey and other educational writers. It was developed into an educational method since late 1960s in the discipline of medicine at McMaster University, Canada (Barrows 2000). Since then it was applied and developed in different educational contexts in great diversity.

consequently the task of the teacher is altered from the transferring of knowledge into facilitating to learn' (Graaff, 1994).

The PBL concept has various definitions and ways in application, ranging from problem-oriented lectures to completely open experiential learning environment aiming at improving interpersonal relations (Kolmos, 1996). With a pragmatic origin and orientation in development, problem-based learning is not a package method with single face, but rather contextualized. Among different definitions and practices, three major characteristics can be distinguished (Graaff and Kolmos, 2003): central theoretical learning principles, specific educational models based on PBL principles, different practices within the guidelines of traditional education models. In brief, PBL in general refers to theory, models and practices (Graaff and Kolmos, 2003).

The concept of problem-based learning takes its theoretical departure from cognitive constructivist and sociocultural theories. There is an assumption that learning is a product of both cognitive and social interactions in a problem-centered environment (Schmidt and Moust, 2000). Working in PBL, learners need to develop Metacognitive strategies to identify what they do not know and what they need to learn more about to solve the problem. They also need to figure out what resources they will need to make up their knowledge shortage. Equally essential is to know how to evaluate the new knowledge in order to know whether it is appropriate and to integrate the new knowledge with prior knowledge to solve the problem.

The practice of problem-based learning varies from country to country, from institution to institution (Kolmos and Vinther, 2004; Jensen and Baekkelund, 2004; Du et al., 2009). Some universities transformed partly or wholly into PBL model, some universities only employ some concepts, some universities employ PBL concept since their establishment.

In the past two decades, PBL has received a large amount of research attention: with more focuses on knowledge acquisition as well as problem-solving advantages compared with traditional learning since the late 1980s, and with a growing interest in self-directed learning or group interactions since late 1990s (Hmelo and Evensen, 2000). Diverse studies have been carried out on different variables[5] of this approach in both psychological and social

[5]Barrows (2000) give some examples of these variables that are often researched: the design of problem formats, the emphasis on teaching problem solving, the training and role of the facilitator, the composition fn responsibilities of the small group members, the process and sequence of learning activities, the

dimensions. In general, these findings concern the understanding of one's own knowledge needs, application of knowledge to novel problem situations, collaboration, and lifelong learning. Some distinguished examples can be witnessed in medical education in both European and North American contexts. In problem-based learning environment, the curriculum provides advantages of the learners' cognitive processes and motivations (Schmidt and Moust, 2000). PBL students are able to solve problems based on progressively more complex principles whereas traditionally educated students could not go beyond the initial context (Barrows, 1985, 2000). Students are more satisfied with the problem-based learning environment than the traditional learning environment (Hmelo and Evensen, 2000). Students see more relevance between what they study and its application, and they are more confident in their problem-solving skills (Vernon and Blake, 1993). Hmelo and Lin (2000) find that PBL students are more motivated to develop effective self-directed learning strategies, and have more awareness of transferring these strategies to new problems and are more effective in integrating their new learning into their problem solving. In the PBL environment, students tend to learn to become good collaborators (Duek, 2000; Koschmann et al., 2000; Du, 2006a).

In general, positive evidences have been found specifically in profession-based education (Hmelo and Evensen, 2000; Kolmos and Graaff, 2007; Du et al., 2009). The above mentioned examples are mostly conducted in the medical discipline. PBL is also widely regarded as a successful and innovative method for engineering education (Graaff et al., 2001; Graaff and Kolmos, 2003; Du et al., 2009). In the following, I will discuss about the practice of PBL as a learning model at Aalborg University.

2.2.2 Problem-based Learning in the Danish Context — PBL Aalborg Model

In Scandinavian countries, through the 1970s and 1980s, PBL was developed on the basis of practical experiences and experiments. It was not until 1990s that serious attention was directed towards the theoretical foundation of PBL, and that the diversity of PBL was researched (Kolmos, 2002). In Denmark,

problem-solving and self-directed learning methods, the use of reflection and assessments, and the degree of student-centeredness.

PBL tradition started from two young universities,[6] in which PBL concepts are practiced somewhat differently. In this study, I mainly refer to the PBL Aalborg Model.

Learning principles

At Aalborg University, the PBL tradition has its theoretical root in the experiential learning primarily developed by Dewey, along with Negt/Kluge's theories of the development of work education and the development of political consciousness formulated at the beginning of the 1970's (Graaff and Kolmos, 2003). More recently, researchers have related PBL concepts to a variety of theoretical notions such as experiential learning by Kolb (1984), constructivism in education, cognitivism by Piaget and Vygotsky, and situated learning by Lave and Wenger (1991). Kolmos (2002) summarizes the central theoretical learning principles and practices that have had significant impact on the development of the problem-based learning model as listed in the following.

- Problem-oriented learning, in which a problem situation provides the starting point and direction for the learning processes. Problem-orientation[7] is defined on the basis of the psychology of learning and originally has a clearly specified political agenda aiming at exposing the structure of the society. In the present practice it serves to promote the motivation and comprehension of learning, because this relates the learning contents to a broader perspective and context.
- Participant-directed and experience-based learning, in which students are expected to find the problem formulations on their own within a given subject framework. The learning process is built from the interests and experiences of students.

[6] In Denmark, PBL was firstly implemented at two universities, Roskilde University in 1973 and Aalborg University in 1974 (Kolmos, 1992).

[7] In the Danish context, problem-orientation has its foundation in a psychological explanation as well as a clearly specified political agenda aiming at exposing the structures of the society. There are different ways of defining problem-orientation and problem-based learning. In many cases, they are used in similar ways since they both rest upon a foundation of constructivist learning theories and are both implemented in higher educational institutions(Jensen and Jensen, 2004). However, some distinct differences are pointed out by Jensen and Baekkelund (2004). Problem-orientation in the context of AAU is closely linked with project work. It emphasises that the problem is identified and formulated by students based on their experiences, but not a subject matter or in the curriculum. In the problem-based learning, the problem is not formulated by the students, but can rather be defined as a task.

- Interdisciplinary learning, in which problems students work with can extend beyond traditional subject-related boundaries and methods and reach real problems and situations.
- Exemplar practice, in which there is the premise that the educational benefit for the student should exemplify the objectives of the framework provided. This means that students must gain a deeper understanding of the selected complex problem. Due to the lack of an sufficiently broad overview of the subject, students are expected to acquire the ability to transfer knowledge, theory and methods from previously learned areas to new ones.
- Group-based learning, in which the majority of learning process takes place in groups or teams. Individual competences are expected to develop along the way of collaboration and cooperation.

In general, these learning principles place learning as the focus of education and take both cognitive and affective parts of learning into consideration. They also go in tune with the ongoing discussion of shedding light on the meaning-seeking and managing personal transformation in the development of engineering education.

Practices

The practice of PBL can be elaborated in the aspects of knowledge generation, curriculum design, problem-solving as learning orientation, project work in groups as learning method, roles of teaching, and assessment.

In PBL Aalborg Model, students are expected to take the responsibility of their own learning and as learners they are expected to participate in the knowledge creation process in stead of only being knowledge receiver (Kolmos, 1991, 2002). In practice, half of the work hours of students are occupied by traditional lectures and seminars, whereas the other half are spent on doing project, which is organized in the form of group work. The size of the group varies and becomes smaller from 6–7 students in the first semester to 2–3 in the final semesters before the completion of a master degree. When doing project work in groups, students will be facilitated with a project room and a supervisor to help them with the project process.

At AAU, problem orientation as one focus in the curriculum is based on working with unsolved, relevant and current problems from society and real life. It emphasises the students' own process of identifying and formulating a problem within a broad framework of reference in the study program (Jensen and Baekkelund, 2004:287). By analyzing the problems in depth, the students learn and use the disciplines and theories that are considered necessary to solve the problem posed, which provides a good balance between theory and practice. Organizing project work allows groups of students to form and structure the whole process from problem formulation, analysis, theories, synthesis, possible solutions and experiments to conclusions, assessment and consequences. Students gain experiences from managing their own project and group work, and they learn how to communicate their ideas, thoughts and meanings to supervisors and other students. This process helps to cultivate their responsibility and cooperative and communicative capability. Other personal qualifications are also obtained in the process of report writing and presentation making (Dresling, 2001).

The employment of project work as the major method of organizing curricula requires a different cognitive style of learning from the one used in traditional disciplinary classrooms (Kolmos, 1992). Project work covers many types of cognitive processes.[8] In this study, project work is used as part of the concept of problem-based and group work organized project work. Specifically, at AAU, project work is defined as a teaching method and pedagogical form (Jensen and Baekkelund, 2004). It functions as a link between theory and practical experiences, which leads to the generation of new knowledge (Kolmos, 1992). It provides students chances to learn to form and structure a whole process from problem-formulation, analysis, theories, synthesis, possible solutions and experiments to conclusions, assessment and consequences (Kjaersdam, 2002).

Kolmos (1996) writes about the differences between project work and assignment in the lecture-based learning environment. In the process of working on assignment, the teacher formulates the questions, knows the methods

[8] Project work as a concept varies in the way it is used. For example, it can be used as the method of part of a single subject, an introductory programme, and fundamental organized curricula. In form, it can be a problem-based project starting by analyzing and formulating the problem, it can also be a case study which is used to illustrate particular elements in specific disciplines. In the way of organization, it can be individual work or group work. (Kolmos, 1992) Illeris (1974, see Jensen and Baekkelund, 2004) extracts three basic characteristics of the term 'project work': work/job oriented, integrating theory and practice, oriented towards society. Jensen and Baekkelund (2004) specify two more points: interdisciplinary and research character.

and has an answer. Therefore, it is a controlled process for the teacher and a directed process for the students. However, in the project work, the teacher might not know the exact problem that is going to be analyzed and solved in the projects. Neither might the teacher know the answers or the outcome beforehand. Nevertheless, they know different methods that can be introduced to the students. For the students, they do not know the direction for the project before they have analyzed and formulated the exact problem, and they need learn methods to solve the problems. In this way, both teachers and students are enrolled in an active and interactive investigating learning process.

The problem-based project work involves a complicated and challenging cognitive as well as affective process. This process becomes more challenging in a group-organized way because learners confront more aspects like feelings, attributes, values, and identity work (Kolmos, 1992). Working in groups demands a variety of other abilities than disciplined knowledge, for example, effective cooperation, positive communication, management, reflection, self-evaluation, which are referred to as process competences (Kolmos, 2002a; Kofoed et al., 2004).

In order to facilitate students with process competences to maximize learning in doing group work, in the first year, a course named Collaborative Learning Project (CLP) is provided to introduce two main landmarks: the organization of the working process in group work and the psychological aspects in collaborations (Kolmos, 1992; Kofoed et al., 2004). At the organizational level, students are given advice on how to make plans to manage the project, how to distribute work, how to discuss individual contribution and put them into a collective project process, how to edit all the contribution in the end in order to submit a common report on basis of the project work. At the psychological level, students are given advice on how to give and receive criticism without any personal attack, how to handle disagreement and conflicts, how to develop individual and common group identities.

Each project group will be facilitated with one or two supervisors. The teaching role as supervisor to the groups is different from that in a traditional lecture-based environment (Kolmos, 2002a; Hohle, 2004). Instead of giving answers and controlling the learning process, supervisors play a role of supporting and guiding students (Hohle, 2004). Supervisors are supposed to be open-minded by leaving room for students to be active, and to ask questions overcoming the feeling of being stupid. They are also expected to give advice for students to develop positive communication and management in the social processes (Kolmos, 2002).

Studying engineering in the PBL environment, doing project in groups with the aim of solving problems functions as the core and thread of learning process. Different from studying in traditional learning environment, where the teacher formulates the questions, knows the methods and provides an answer, working on projects involves a process of active learning. None of the supervisors or students knows the exact problem which will be analyzed and solved in the project. Supervisors might know different methods that can be used, but nobody know the result or outcome beforehand. Students, rather than supervisors, are responsible for the project work.

Assessment in this learning model also varies based on the design of courses, which generally fall into two types. One type of course is relevant to the project work. This kind of courses will be assessed through the examination of their shared project work, which is evaluated in forms of project report and oral defense as a group. Oral defense will be charged by a main examiner, which is normally the supervisor, together with a censor from another institute or school. The other type of courses is not directly relevant to the project work, which will be assessed individually in forms of oral or written exams.

Evaluation of learning

A variety of investigation has been conducted with the purpose of evaluating the learning outcome of studying engineering in PBL Aalborg Model. The target groups include undergraduate students, supervisors, external examiners, graduates, and employers. Assessments from educators' perspective conclude that the PBL environment equips engineering students with analytical skills, capabilities of solving problem and applying theories to practice,openness towards new knowledge, and awareness of active and responsible learning (Kolmos, 1996; Kjarsdam, 1992, 1994; Fink, 2001). Surveys from the employers' points of view show that students educated from PBL environment acquire better competences in collaboration, management and entrepreneurship (Kolmos, 1992; Kjarsdam, 2002). Investigation from the students' perspective indicates high level of motivation, reflection on learning and satisfaction with learning(Kolmos, 1992; Kjarsdam and Enemark, 1994; Kjarsdam, 1994a).

A comparative evaluation from international panel (Kjaersdam and Enemark, 1994) witnesses significant differences between graduates from PBL education and those from traditional education. Graduates from AAU are more adaptable to the engineering workplace due to the emphasis on synthesis and group work at university; whereas graduates from traditional education are better at theoretical knowledge and more capable of working independently.

An investigation on the study strategies of first year master students at a traditional learning environment from 1999 to 2002 provides evidences of mismatch between learning and teaching (Christensen, 2002). The main reason leading to this mismatch, as analyzed by Christensen (2002) is that due to the lecture-based teaching tradition, students adapt study strategies suitable to meet the requirement of the teaching method and grant good grades, which does not necessarily lead to deep understanding and learning. The lack of dialogue between teachers and learners results in mismatch in expectations as well as objectives between teaching and learning. In practice, students spend almost half of their study time on theoretical calculations and little time on real problems. As concluded from this research, only changing teaching methods might activate students' response to teaching and examination, but attending lectures does not significantly increase independent learning.

It has been argued that evaluation system in engineering education in general prioritizes the cognitive part of learning and the mastery of the expected engineering skills (Salminen–Karlsson, 1999). The dependence of learning outcome on the context in which learning takes place has been recognized as an important factor in terms of improving the quality of engineering education (Kolmos, 1991, 2002; Christensen, 2004). In the PBL context, given that theoretical literature suggested that learning is not an accumulation of information, but a transformation of the individual who is moving toward full membership in the professional community (Hmelo and Evensen, 2000). In practice, the group organized project work challenge the development of ones' personal identity and that they have to work on the development of this identity in a group context. Therefore, more attention is needed on the identity making process through learning as well as the sociocultural context of PBL, which are equally important components of the curriculum (Kolmos, 1992; Hmelo and Evensen, 2000; Du and Kolmos, 2009).

2.3 Gender Issues in Engineering Education

This section brings about discussion on another aspect of diversity issue in engineering education, that is, to make it attractive to different social groups and to introduce values of different social groups into the development of engineering education. From a feminist perspective, I take woman as a social group for analysis in this study.

2.3.1 The Historical Underrepresentation of Women

Engineering has been historically associated with mechanical activity and creation of technology, and therefore, it has been closely connected with male sphere in the western society (Cockburn, 1985; Hacker, 1989; Kvande, 1999; Salminen–Karlsson, 1999). From a gender perspective, this close connection between engineering and masculinity has been regarded as the main reason for the historical exclusion of women from engineering education by feminist scholars from many cultures in the world.[9] These feminist analyses in general fall into three aspects.

Firstly, historical studies on engineering education have recognized some reasons that make engineering education distant from women. In this profession-oriented education, students are provided with engineering expertise by being trained to think and work in particular ways. Technological schools in western countries have a historical ideology of turning out effective, rational, well-disciplined engineers, which is closely linked with male gender ideology in western society (Connell, 1995) and male homosociality that is dominating the engineering profession (Hacker, 1989; Cockburn, 1991; Berner and Mellstrom, 1997; Hodgkinson and Hamill, 2006). These characteristics have defined engineering as an elite community, made its members a homogeneous group (a small amount of men), and led to the historical exclusion of some social groups, for example, women (Hacker, 1989; Berner and Mellstrom, 1997; Kvande, 1999; Salminen-Karlsson, 1999; Powell et al., 2009; Riley et al., 2009).

[9] Engineering as a male-dominated place, where women are marginalized has been recognized by feminist scholars in different places in the world, such as, in Scandinavian countries (Kvande and Rasmussen, 1994; Kvande, 1999; Brandell, 1996; Brandell, et al., 1998; Berner and Mellstrom, 1997; Edelman, 1997; Salminen–Karlsson, 1999, 2002, forthcoming; Dahms, 1999; Du, 2005, 2006abc; Du and Kolmos, 2009), in other parts of Europe, like the U.K. (Webster, 1997; Henwood, 1996, 1998; Phipps, 2002; Williams, 2002), Portugal (Williams, 2002), the Netherlands (Hermanussen and Booy, 2002), Italy (Gherardi, 1995), Germany (Sagebiel and Dahmen, 2006; Ihsen and Gebauer, 2009), in Australia (Roberts and Ayre, 2002; Male et al., 2009) and New Zealand (Stonyer, 2002), in North America like Canada (Frize, 1993; Dececchi et al., 1998; Dryburgh, 2002), the U.S. (Hacker, 1989, 1990; Fish, 1995, Shull and Weiner, 2002; Tonso, 1996ab, 2007; Franchetti et al., 2010), and in Asian countries (Sim and Hensman, 1994; Du, 2001; Sulaiman and AlMuftah, 2010).

Secondly, previous on-campus research has recognized some implicit reasons that make engineering curriculum indifferent for women. 1) The curriculum content has been strongly based on sophisticated natural science knowledge (like mathematics and physics) and hard-core technology, which is more closely linked with male experiences and male gender socialization. At the same time, in engineering curriculum, there has been a lack of relation to societal implications of technology and involvement of social concern, which is more close to female nature (Kvande, 1999). 2) The harsh engineering culture makes it a tough and challenging experience to get through engineering education, which has been witnessed as a common characteristic in the contexts of Scandinavian cultures (Kvande and Rasmussen, 1994; Berner and Mellstrom, 1997; Salminen–Karlsson, 1999) and other western cultures like the U.K. (Henwood, 1996, 1998), and the North America (Hacker, 1989; Seymour and Hewitt, 1997). The established practices, norms, and values in engineering education have been mainly based on men's experiences and bring about special need and extra identity work for female engineers to adapt to the established male-norm culture (Kvande, 1999; Du, 2006c).

Thirdly, a visible distance between engineering schools and workplaces (in spite of the inherited engineering professional culture) and more harsh culture has been identified in different literature (McIlwee and Robinson, 1992; Kvande and Rasmussen, 1994; Mellstrom, 1995). At engineering universities, the confidence of students is achieved from academic performance, which is based on science subjects like math and physics. In this way, women's academic talents can be recognized and rewarded due to their diligence and strong will in studying science and technology (McIlwee and Robinson, 1992). However, the values in workplaces put more weights on technical proficiency and skills of applying the scientific knowledge to problems-solving. The stress of technology centrally in engineering workplaces involves established practices, for example, technology-related languages, certain interactional and communicational patterns. This distance brings about a gap between being an engineering student at the university and becoming an engineer at the workplace (Du, 2006). In particular, this distance makes it more difficult for women than for men to transfer from educational culture to workplace culture due to their lack of tinkering skills, hands-one experiences, and inappropriate gender social role (McIlwee and Robinson, 1992; Kvande and Rasmussen, 1994; Kvande, 1999; Powell et al., 2008; Sulaiman and Almuftah, 2010).

As can be concluded, engineering education are attractive to only a certain group of young men for the following reasons: 1) it has historically provided greater privileges for men than for women; 2) the current practices, which is dominated by male interest and hardcore technology; 3) the homogeneous male-norm culture in engineering workplace. As Seymour and Hewitt (1997) comment, the covert function of engineering education is not beneficial neither for male students nor for female students, but in particular, it builds up barriers for women to study engineering. In general, it can be said that engineering education has been and is still created and conducted mainly by men, and that it shows features which please and privilege men while trouble and disadvantage women (Salminen–Karlsson, 1999).

2.3.2 Gender Issues in the Danish Context

In Denmark, women's participation into many educational settings has been growing rapidly in recent decades. The increase in women's participation has exceeded that of men's, and there are currently more women than men receiving Bachelor's and Master's education (Dahms, 1999). Although more and more women have access to higher education and women even constitute the majority of students in higher education, there continues to be visible patterns concerning the subjects they choose to study. Women are more clustered in traditionally female fields and it is difficult to find women's presence in male-dominated area, engineering, for example (Dahms, 1999).

Women have been under represented in engineering education in Denmark in that engineering has historically been a non-traditional profession for women (Dahms, 1998, 1999). According to the interview with a Danish historian[10] on engineering education, women's absence in engineering has seldom been addressed in research on the history of engineering education. During my research, there has been limited works that can be turned to regarding research experiences in gender and engineering education.

In 1980s, some initiatives have been taken to call for gender equality in the areas of science and technology. A number of corresponding activities like awareness-raising seminars and conferences, recruitment campaigns, information materials and so on led to an increase of women's participation

[10] see Chapter 6.

into engineering education from below 10% in 1980 to around 25% in 1990 (Dahms, 1999). However, actions during this period were mainly in the form of information providing and recruitment campaigns with few attempts of changing study forms and curriculum contents. They were focused on 'changing the women and bringing them into male dominated education and jobs' rather than changing the norms and values in engineering profession at large (Dahms, 1998, 1999). But these women faced severer difficulties in job hunting after graduation, since unemployment among engineers increased from 2.5% 1985 to 9% in 1990, but with female unemployment rates being twice as high as male, which resulted in the drop of female participation in engineering education to around 17% in 1995 (Dahms, 1999).

Since the mid-1990s, the emphasis of feminist work has been turned to the social structure by questioning engineering education and workplaces instead of women. Different efforts have been made with an aim of recruiting more women into engineering, for example, establishing women's committees in the Engineering Educational Co-operative Council (IUS),[11] and the Danish Engineers Society (IDA),[12] carrying out mentorship programs,[13] introducing new study programs. However, no visible success in terms of numbers has been witnessed.

2.3.3 Changing the Education Instead of Women

The last decade in 20th century witnessed substantial efforts in different western countries aiming at women students to science and engineering-related studies through policy development and attempts of curriculum reforms, as exemplified in Chapter 1. The expectation that these would increase the appeal of higher technical studies for girls has partly been met. However, these educational reforms have alone not been the key to a larger intake, retention and output of girls (Hermanussen and Booy, 2002). Many institutions have experienced that the measures taken until then had been insufficient to bridge the

[11] The Engineering Educational Co-operative Council (IUS) has established a Women's Committee in 1996 with the aim of looking at means and ways of attracting young women to engineering. But this committee has dissolved itself because of failure in fund raising efforts (Dahms, 1997).

[12] A Women's Committee has been established in the Danish Engineers Society (IDA), formulating and recommending policy concerning women and equity in IDA.

[13] The mentorship programs were launched in AAU since 1998 to provide women engineering students with voluntary women engineers as mentors, helping them with all the problems facing in the university period with the aim to reduce the dropout rate (Dahms 1999).

large gap between girls and technology. As Salminen–Karlsson concludes (1999, 2002), increasing the percentage of women in enrollment does not automatically solve the problems of the male hegemony of the education; to really make the engineering education female-friendly, a bigger social change is needed — all the elements like curriculum content, teaching methods and the prevailing culture in engineering should be changed.

2.4 A 'Friendly' Learning Environment for Gender Diversity in Engineering Education

Based on what has been discussed about the trend of making changes in engineering education and the application of PBL concept to the educational practice, this section assumes that a friendly learning environment can be an option to increase diversity in engineering education in terms of increasing diverse competences and attracting diverse social groups (women in particular in this study). Through discussing criteria for a friendly environment, this section aims to establish a basic frame to examine the research context (PBL environment in engineering education) in my research.

2.4.1 Learning Environment in Engineering Education

Recent years witnessed more and more attention regarding how to provide a friendly learning environment in order to improve the quality of engineering education and provide qualified graduates as future engineers (Christensen, 2004; Kolmos et al., 2004; Du and Kolmos, 2009). Learning environment is an indistinct concept and can refer to physical environment or cultural environment. Christensen (2004) mentions three mutually depending and interacting elements for a learning environment: studying (which is the main task of students as learners), facilitating (which cover different teaching activities) and framing (which refers to context that includes both physical surroundings and mental conditions). According to Christensen (2004), a prescribed supportive learning environment provides both formal and informal activities in order to maximize learning opportunities for learners.

The design of a learning environment depends on the competencies the learning should provide to the student (Christensen, 2004). In relation to engineering education, the aim of establishing a learning environment should be linked with the expected engineering competences. Based on the above-mentioned engineering competencies, a learning environment for engineering students should be should not only support deep understanding and engineering knowledge, but also provide learning skills by encouraging engagement in self-directed[14] learning, active learning and cooperative learning (Kolmos, 1991, 2002; Christensen, 2004).

2.4.2 Evaluating a Learning Environment from a Gender Perspective

From a feminist perspective, Rosser (1996) and Lewis (1993) elaborate different aspects that concern what is required for technical and scientific education to be gender inclusive. Rosser (1996: 245) specifies some stages that curriculum reforms need go through to lead to gender equality in science education.

1) absence of women not noted;
2) recognition that science has a masculine perspective;
3) identification of barriers to women in science;
4) search for women scientists and their unique contributions;
5) analyzing science done by feminists/women;
6) Science redefined and reconstructed to include us all.

These stages, according to Rosser (1996) do not have to follow in the above-mentioned order, and they can be used to evaluate a curricular reform in science education. These stages can be used to aid evaluation of engineering education with a specific aim of calling for gender awareness in engineering education.

Through addressing some commonly identified issues for female students in male dominated courses, Lewis (1993) pointed out three aspects that have been neglected in the construction of science and engineering curriculum. They are respectively 1) the construction of the curriculum with the consideration of the students' background in light of formal and informal experiences and interests; 2) student/student interactions; 3) teacher/student interaction. Lewis attributes these issues as main reasons that made the traditional science and

[14] Zimmerman and Lebeau (2000) suggest six indicators for the self-directed learning. 1) Defining what should be learned; 2) Identifying ones' own learning needs; 3) Developing learning objectives; 4) Identifying a learning plan to achieve those objects; 5) implementing the learning plan; 6) Self-evaluating the effectiveness of the learning.

engineering curriculum female exclusive, and asserts that real changes need to be made in all these aspects in order to make the curriculum female friendly.

The criteria proposed by both of the writers go in tune with the advocating of changes in three aspects that came out of the European Curriculum-Women-Technology (CuWaT, 1998). This project proposed a three-dimensional approach to make changes in engineering curriculum to make it gender friendly. These three dimensions are respectively: 1) changes in pedagogical approach, which include the shift from teaching to learning as the focus of education, changes of the roles of students and staff in education, and changes of classroom climate or learning environment; 2) changes in contents to assure the conceptuality of knowledge; 3) changes in learning culture in different engineering programs.

2.4.3 Is a PBL Environment for Gender Diversity?

In general, the PBL concept at AAU has been evaluated as an effective educational system in training future engineers who are prepared for the challenges of the globalization, and who can meet the requirements of the profession competencies. It has also been considered as a supportive environment for studying engineering from both cognitive and affective perspectives. From a gender perspective, some feminist scholars have argued (Hayes, 2000; Fox et al., 2009) that women's learning is better supported in an environment that is different from those in traditional higher education and from those that support men's learning. It has been believed that some characteristics of the PBL environment like the expectation to be cooperative, communicative and its association with some social, sociological, cultural, and environmental concern are supposed to support women's learning (Kolmos, 1991; Du, 2003, 2006b,c; Du and Kolmos, 2009; Fox et al., 2009). However, there has been little empirical evidence in this aspect.

The assumption that a PBL environment in engineering education might be a friendly learning environment to learners of both genders establishes the main research objective of this study. Based on what has been discussed in this section, this study regards a 'friendly' learning environment as a milieu that supports both cognitive and affective parts of learning. Accordingly, a 'gender' friendly learning environment is a milieu that supports the learning of both men and women. Taking a departure from the learner's perspective, this study places a special focus on the perception of learning from engineering students of both genders as well as their appreciation of meaningfulness of learning and identity work process through studying engineering in a PBL environment.

Summary

In this chapter I have addressed the need for increasing the diversity in engineering education, which focuses on two aspects: diverse competences that future engineers need to be equipped with and diverse social groups (for example, women as a social group that has been historically excluded from engineering) that will show interests in studying engineering and that will bring diverse values and contributions into the development of engineering education. In order to achieve these goals, a friendly learning environment is assumed as a positive option. Through providing the basic information about the PBL environment as a learner-centred environment in the Danish engineering education and relating a gender perspective to this discussion, I hope that I have made the cultural context of this research understandable.

The choice of this research context as well as one of the objectives of this research, that is, to examine the practice of PBL concept in engineering education from the perspective of students, to some extent leads to the choice of drawing on theories of learning from constructivist-sociocultural perspective (in Chapter 3). The background of the gender issues in engineering education helped with the choice of relevant gender theories (in Chapter 4). Based on this context, in the following chapters, I will examine learning theories and gender theories and establish a theoretical framework for the analysis of the empirical data.

3

Understanding Learning

This chapter starts with an introduction to the choice of constructivist-sociocultural approach as the theoretical standing point in concern with understanding learning in this study. This is followed by a brief review of literature which examines learning from different perspectives such as experiences, situation/context, participation, and identity development. A number of learning theories in the constructivist-sociocultural approach are discussed and linked with a philosophical concept 'Bildung'. Based on this discussion, a method of analysis combining the concept of 'Bildung' and the constructivist-sociocultural approach is proposed through building up a model for understanding learning from three levels, individual level, institutional/community level, and sociocultural level. Central to this model of understanding learning is the highlight of meanings seeking and identity development. It is also an attempt to establish an analytical tool to relate to the field of learning and education (in particular, the research context of engineering education in this study) for the examination of the empirical work.

3.1 A Choice of Theoretical Departure to Examine Learning

There are various approaches exploring theories of learning in terms of what learning is and how it happens. Psychologists, anthropologists, linguists, philosophers and educators are trying to understand how the mind works and how people learn from different perspectives. Therefore, there remain

Gender and Diversity in a Problem and Project Based Learning Environment, 31–55.

diverse beliefs and definitions on learning. It is a difficult task to distinguish these different approaches clearly since each approach is undergoing changes throughout human history. These changes also bring about various overlaps regarding perceptions of the nature of knowledge as well as understanding of learning both theoretically and empirically.

In this section, I will have a brief review of a number of approaches that have been well-employed in the history of western educational settings, for example, behaviourist approach, cognitivist approach, and construction-sociocultural approach. The intention of this review is not for categorization, but rather, to gain an overview of different ways of understanding knowledge and exploring learning. Through discussing the relation of each approach to my research, I will provide a theoretical rationale for employing a constructivist-sociocultural approach to examine learning. The point of departure for this choice is from the learning principles of the PBL environment (as discussed in Chapter 2) as well as my research interest in the meaningfulness in learning.

3.1.1 Behaviourist Approach

Emerging after the Enlightenment, behaviourism focuses on objectively observable behaviours and discounts mental activities. For this approach, knowledge is a collection of information, and it is more explicit rather than tacit (Bigge and Shermis, 1999). Accordingly, Learning is seen as a change in observable behaviour, and it is often defined as the acquisition of new behaviour (Woolfolk, 1987; Jarvis et al., 1998). Behaviourists believe that through stimuli (the cause of learning, for example, environmental agents) and responses (effects of learning, for example, physical reaction), learning can be controlled to produce a desired effect.

Behaviourist theory has been widely accepted and practised in education, when schools expect to get measurable results from students (Phillips and Slotis, 1998). It is often used in schools where teachers are the only resource of information and knowledge, and students are expected to take in the expertise passed on by teachers. Good students do as they are told and conform to accepted norms. Rewards and punishments are often-used strategies to stimulate expected behaviour and required response from students.

Behaviourist theory on learning is criticized in many aspects (Woolfolk, 1987; Jarvis et al., 1998). Firstly, it only focuses on the measurable behavioural outcomes of learning rather than knowledge, attitudes, values, beliefs and so on. Secondly, by focusing on the result of learning, which is evaluated by the sum of human behaviour, behaviourism misses the understanding of learning processes. Thirdly, behaviourism put too much emphasis on the one-way influence from the learning environment to the individuals. However, due to the concern about end products and their assessment, behaviourist theories of learning still dominate some branches of teaching methods as well as some research on learning (Jarvis, 2001).

The beliefs of behaviourism on learning are rather different from the principles of the PBL learning environment in my study, even though some hints of this approach might be witnessed in some teaching practice in reality. There has been paid little attention to the meanings of learning in the works of this approach. Therefore, this approach only functions as background knowledge in this study.

3.1.2 Cognitivist Approach

Cognitivism is known as an information processing theory which deals with how people perceive, learn, remember, and think about information (Woolfolk, 1987). It believes that people learn from past experiences and learn by thinking. As a process of gaining or changing insights, outlooks, expectations or thought pattern, learning is assumed to be internal, and involves recognizing, interpreting experiences and understanding the meanings.

In education, cognitivist teachers play a role of coach and encourager (Phillips and Slotis, 1998). They aspire to help students develop their understandings of significant problems and situations. They seek to provide an effective learning environment which includes models to imitate, meaningful activities, progressively difficult tasks and diverse problems to solve, and so forth.

Cognitivist approach of perceiving learning is criticized for regarding knowledge as only facts and by assuming that learners are passive receivers of knowledge (Woolfolk, 1987). Another critique concerns the assumption that knowledge can be acquired independently of their contexts for use so that people can predictably transfer learning from one situation to another (Jarvis et al., 1998).

Jean Piaget is well-known as one of the most contributing cognitivism theorists. Piaget's works has a great influence on the general understanding of individuals' development. His theoretical framework 'genetic epistemology' is originated from his primary interest in the knowledge development in human organisms, that is, how we come to know (Huitt and Hummel, 2003). This theory is centred on concept of cognitive structure. Cognitive structures are patterns of physical or mental action that underlie specific acts of intelligence and correspond to stages of child development. Cognitive structures change through the processes of adaptation: assimilation and accommodation. Assimilation involves the interpretation of events in terms of existing cognitive structures whereas accommodation refers to changing the cognitive structure to make senses of the environment. Cognitive development consists of a constant effort to adapt to the environment in terms of assimilation and accommodation.

Piaget's works are applied extensively to teaching practice and curriculum design. To Piaget (Huitt and Hummel, 2003), educators should be encouraged to plan a developmentally appropriate curriculum that enhances the logical and conceptual growth of students. Teachers need emphasize the role of experiences or interactions with the surrounding environment on the learning process of students. In the PBL context, Piaget's ideas are well-employed in the teaching practices (Kolmos, 2002), especially in the aspect of developing supervision skills.

The cognitive approach of understanding learning has been an effective tool to examine the individual learning process. However, it fails to take the affective part of learning (such as emotions, attitudes, values, etc.) into consideration. Neither does it account the interaction between individuals and the specific social and cultural context. The overemphasis on perceiving learning as a pure individual activity makes it difficult to answer the questions about interaction between individuals and structure in this study (for example research questions 2 and 3).

3.1.3 Constructivist-sociocultural Approach

A common belief shared by behaviourist approach and cognitivist approach is that learning is merely an individual activity. Learners, according to these theories, might either actively interact with the physical environment or passively receive the simulation or experience. However, what is missing from all these accounts is the explicit recognition that 'learners always belong to social groups' (Phillips and Soltis, 1997:53). Different from behaviourist and cognitivist approaches of focusing on the examination of how individuals learn, constructivist and sociocultural theorists are devoted to examining the social dimension or context for learning.

Inspired by Piaget's theory about how young learners construct their knowledge structures, constructivist approach was developed as a domain of contemporary movement in education. Constructivism is a philosophy of learning founded on the premise that by reflecting on experiences, people construct their own understanding of the world they live in. Everybody makes sense of their own experiences by generating their own mental models and rules; therefore, learning is perceived as a process of adjusting the mental models to accommodate new experiences (Jarvis et al., 1998).

For constructivist educators, learning is a search for meaning, rather than just memorizing the 'right' answers and repeating someone else's meaning. Thus, they call for the elimination of a standardized curriculum. Constructivists emphasize learning by doing and hands-on problem solving. They encourage students to analyze, interpret and predict information and support students to foster new understanding based on past experiences.

A well-known representative of this approach, Philosopher and educationist John Dewey is very sensitive on the social nature of learning. As he notes in 'Democracy and Education' (1916):

> As matter of fact every individual has grown up, and always must grow up, in a social medium. His responses grow intelligent, or gain meaning, simply because he lives and acts in a medium of accepted meanings and values. Through social intercourse, through sharing in the activities embodying beliefs, he gradually acquires a mind of his own.

Dewey developed a 'problem-solving' theory of learning whose basic premise was that learning happens as a result of our 'doing' and 'experiencing' things in the world as we successfully solve real problems that are genuinely meaningful to us (Phillips and Soltis, 1998). Dewey described schools as communities. School learning, according to Dewey (1938), should be based on meaningful student experiences and genuine student problem solving. He believed that school learning should be an experientially active, but not passive affair. Learning take place mostly through communication with others, which refers to purposeful interaction with teachers and fellow students. Based on these beliefs, Dewey suggested that students should be engaged in meaningful activities where they need work on problems with others.

The Sociocultural approach has its root in the belief that cognition is intertwined with other people, tools, symbols and processes (Vygotsky, 1978; Wertsch, 1985; Lave, 1997; Rogoff, 1995; Wenger, 1998). The analysis of this approach is focused on people's engagement in activities and the construction of interactive understanding. Socioculturalists view learning as activities of individuals' participation in cultural practices and interaction with others in the social contexts.

Vygotsky (1978) considered as the founder of sociocultural theory, developed a comprehensive framework that covers individual cognition, social institutions and cultural meanings. Specifically, in his theory of 'zones of proximal development' (ZPD), Vygotsky suggests that the development of individuals depends on the interaction with people and culture, which provides a tool for people to form their own view of the world.

In relation to education, Vygotsky believes that curricula should be designed to emphasize interaction between learners and learning tasks (Bigge and Shermis, 1999). Scaffolding — where the instructor continually adjusts the level of help in response to the learner's level of performance — is an effective form of teaching. Vygotsky suggests three ways in which a cultural tool can be passed from one individual to another (Bigge and Shermis, 1999), which are respectively imitative learning, instructed learning, collaborative learning (which involves a group of peers who strive to understand each other and work together to learning a specific skill).

Both constructivist and sociocultural approaches are playing increasingly prominent roles in the educational development in the past decades (Cobb and Yackel, 1996; Phillips and Soltis, 1997; Jarvis et al., 1998; Bigge and Shermis, 1999). In practice, they are different in many ways regarding the focuses on analysis. For example, constructivists have more emphasis on how individuals construct meanings through concrete practice, while socioculturalists stress more on the cultural practices and social environments in which individuals participate. However, in my study, focus is put on their shared beliefs on learning, that is, learning is constructive rather than productive, and learning is primarily a social, cultural and interpersonal process. Therefore, these two perspectives are seen as one approach in my study for the analysis of learning as an individual-institutional-cultural interaction.

The constructivist-sociocultural approach has a close link with the PBL environment. Established its foundation under the constructivism umbrella, some learning principles of the PBL environment, for example, problem-solving and collaborative learning, have their origin from both Dewey's and Vygotsky's works (Kolmos, 2002). This, together with the basic beliefs of learning held by this approach, serves as the main reasons for taking this approaching as a theoretical departure of my research.

3.2 A Review of Relevant Works in Constructivist-sociocultural Approach

The constructivist-sociocultural approach plays an influential and foundational role on the work of many scholars later. For example, Kolb's work (1984) on experiential learning is partly developed from the works of both Piaget and Dewey. The work of Vygotsky and Dewey has inspired a number of scholars to develop further the idea that human thinking and learning is not a process that only involve the inside of the human cranium. Based on these thoughts, Lave and Wenger (1991), Wenger (1998, 2009), Lave (2009) develop their work on situated learning and communities of practice. Inspired by Vygotsky's focus on three themes (that is individual, interpersonal and sociocultural), Rogoff (1994, 1995, 1997) developed her participation theory to analyze learning activities. In this section, these different works will be discussed in more details and be related to the learning principles of the PBL environment in my research.

3.2.1 Learning from Experiences

Since Dewey wrote Experience and Education (1938), recognizing how experiences lead to learning, learning from experience has become an agreed key feature of much contemporary thinking about adult learning. Experiences play a central role in the learning process from the perspective of experiential learning theorists in understanding the function of education, the proper relationships among learning, work, and other life activities, and the creation of knowledge itself (Jarvis et al., 1998). Among different writers, David Kolb's work[1] (1984) plays an influential role in the discussion about experiential learning. Making explicit use of the work of Piaget, Dewey and Lewin, Kolb lays his interest in exploring the processes associated with making sense of concrete experiences and the different styles of learning that might be involved.

Kolb (1984) describes learning as a holistic adaptive process that 'provides conceptual bridges across life situations such as school and work'. Learning is conceived by Kolb as a continuous process where knowledge is derived from, and tested out, in the experiences of learner, rather than as mere outcomes represented by the amount of fixed ideas that have been accumulated. Learning is thus defined as 'the process whereby knowledge is created through the transformation of experiences' (1984:38). This definition emphasizes on the process of learning rather than the outcome, in turn, knowledge is not an independent entity to be acquired or transmitted, but rather, a transformation process being created and recreated.

Kolb's theory postulates that individuals learn and solve problems by progressing through a four-stage cycle: concrete experience, observation and reflection, the formation of abstract concepts and testing in new situations. Kolb argues that the learning cycle can begin at any one of the stages and it should in reality be approached as a continuous spiral. However, he suggests that the learning process often begins from the learners carrying out a particular action and then seeing the effect of the action in this situation. The first stage **Concrete Experience** corresponds to 'knowledge by acquaintance', which refers to direct practical experience ('apprehension' in Kolb's terms). This is opposed to 'knowledge about' something, which is theoretical, but perhaps more comprehensive, ('comprehension' in Kolb's term). The second step is to understand these effects through observation in the particular distance. **Reflective Observation** concentrates on what the experience means to the learners who have had direct experiences, (it is transformed by 'Intention') or its connotations. In the stage of **Abstract Conceptualization**, learners use logic and ideas rather than feelings to understand problems or situations and the general principles

[1] Kolb's experiential learning theory (1984) has been placed into different categories. Sometimes it is placed in the paradigm of cognitivist approach, sometimes it is placed in the independent paradigm of seeing learning from experiences. In this study, I place works of examining learning from experiences (with Kolb's work as an representative) in the category of constructivist approach due to its origin from Dewey's work.

under which the particular instance falls. They rely on systematic planning and develop theories and ideas to solve problems, and then they transform the theories and hypotheses by testing it in practice (by 'Extension') and relate it to denotations in the stage of **Active Experimentation**.

Kolb's experiential learning theory plays a significant role in locating learners as the centre of learning. Therefore, his learning circle has been widely used in curriculum design for different student-centred learning environment. It also lends an effective tool in the understanding and instructing individual learning in the PBL Aalborg Model as well as in contributing to the further development of this learning environment. In practice, it is used in the staff training activities to help teachers plan their teaching activities and develop strategies to improve the efficiency of the learners' engagement (Kolmos et al., 2004). It is also used in the CLP courses to help students have the awareness of as well as develop practical skills of learning from experiences and from reflecting on the experiences, and how to link theories with practice (Kofoed et al., 2004).

Kolb's learning circle has also received critiques, which mainly come from two directions. Firstly, as Jarvis (1995) and Tennant (1997) comment, even though Kolb has put forward some particular learning styles, the experiential learning model does not apply to all situations. There are alternative situations of learning such as information assimilation, and memorization that have been neglected (Rogers, 1996). Secondly, Kolb is criticized that he only related learning and knowledge in an intimate way through a basic psychology perspective, but failed to explore that nature of knowledge in depth within philosophy and social theory (Jarvis, 1995; Tennant, 1997).

Relating Kolb's learning circle to the research context in my study, I argue that there are some issues regarding the understanding of learning that cannot be explained by the experiential learning circle. (1) Some elements of learning process can not be found in the learning model, for example, setting up the learning goals, choice and decision making, meaning seeking, and identity work. (2) With the focus on knowledge as outcome of the combination of taking in experience and transforming it, learning is more perceived as a passive activity rather than active one. (3) This circle does not explain the function of social context on learning. For example, different situations in the group work, such as gendered distribution, might influence individuals' ways of processing experiences as well as the affirmative part of learning.

3.2.2 Learning as Situated

From the perspective of sociology, some writers see learning as situated activities (Lave, 1988, 2009; Brown et al., 1989; Lave and Wenger, 1991). In contrast with most classroom learning activities that involve knowledge which is abstract and independent of context, these writers suggest that learning is a function of the activity, the context and the culture in which it normally occurs. From this perspective, knowledge is not a thing or set of descriptions, therefore, learning is not only the acquisition of certain forms of knowledge, but rather, learning arises in all human activities in the daily life.

Social interaction is a critical component of situated learning. Brown and Collins and Duguid (1989) point out that learning, both outside and inside school, advances through collaborative social interaction and the social construction of knowledge. Agreeing with them, through introducing the concept of 'legitimate peripheral participation', Lave and Wenger (1991) propose that learning is a process of participation in communities of practice. Participation is at first legitimately peripheral but increases gradually in engagement and complexity. As the beginner or newcomer moves from the periphery of this community to its center, they become more active and engaged within the culture and hence assume the role of expert or old-timer.

Three aspects of situated learning approach have been highlighted in the understanding of learning (Tennant, 1997): (1) the emphasis on the occurrence of learning through relationships; (2) the emphasis on learners' participation; (3) emphasis on the intimate connection between knowledge and activity. These characteristics provide significant pointers for the educational practice in PBL environment Aalborg Model, where the attention has been paid to understand knowledge and learning in context, and where a belief is held that learning will be maximized through learners' participation and collaborative social interaction. In particular, the design of different engineering programs takes the theoretical departure that studying engineering is not merely the acquisition of certain professional technical knowledge, but also a process of active learning activities (Kolmos, 1991, 2002). Therefore, learning will be supported when students are encouraged to acquire, develop and use cognitive tools in authentic domain activities, for example, problem-solving and doing project work in groups.

Situated learning theory is claimed to focus on off-school learning (Lave and Wenger, 1991), in relation to the research context in my study, I find some aspects that are beyond the explanation of this approach. Firstly, when significant weight is placed in context and interpersonal interaction as learning resources, it is difficult to explain how to achieve abstract knowledge and individual cognition, which are regarded as important factors in traditional formal schooling settings. In particular, traditional engineering and science curriculum involves plenty of knowledge that is not necessarily related to daily life situation. As Tenant comments, situated learning theory depends on the claim that 'it makes no sense to talk of knowledge that is decontextualized, abstract or general' (1997:77). Secondly, I argue that, in spite of the emphasis on the knowledge-activity relation in learning, the focus in situated learning approach is restricted to individual activities and does not shed enough lights on interaction between individual and structure at the organizational level. In this way, situated learning theory only looks at learning as one-way receiving knowledge but does not include the generation of new knowledge.

3.2.3 Learning as Participation

Participation as a resource of learning has been addressed in the works of experiential learning and situated learning. In contrast with perceiving learning as internalization, conceiving learning in terms of participation focuses attention on ways in which it is an evolving, continuously renewed set of relations (Lave and Wenger, 1991:49–50). However, none of the writers in those two approaches illustrate 'participation' as an explicit concept on learning. Through employing the work of Rogoff (1994, 1995, 1997), I intend to bring this concept as one perspective into the framework of understanding learning.

Rogoff (1995) proposes a sociocultural approach that involves observation of development in three planes of analysis corresponding to personal, interpersonal, and community processes, which are respectively participatory appropriation, guided participation, and apprenticeship.

Apprenticeship refers to community activity, which involve both active individuals participating with others in culturally organized activity. This concept focuses the specific nature of the activity involved as well as on its relation to practices and institutions of the community in which it occurs. Originated from the idea of craft apprenticeship, this metaphor indicates that participation into activities include the development of people from less experienced to mature status. As Rogoff notes, apprenticeship as a concept goes far beyond expert-novice dyads; it focuses on a system of interpersonal involvements and arrangements in which people engage in culturally organized activities in which apprentices become more responsible participants.

Guided participation refers to the processes and systems of involvement between people as they communicate and coordinate efforts while participating in culturally valued activity. By 'guidance' Rogoff means the direction based on cultural and social values. 'Participation' is refers to observation as well as 'hands-on' involvement in an activity. It includes different indicators that encourage or discourage people, covering from deliberate attempts to instruct as well as incidental comments or actions that are overheard or seen as well as involvement with particular materials and experiences that are available. For Rogoff, participation does not only include face-to-face interaction, but also the frequent side-by-side joint participation in everyday life and more distant unknown people (for example, peer, experts, family, neighbors or distant heroes etc.), or arrangement of activities that do not require co-presence (for example, choices of where and with whom and with what materials and activities a person is involved).

Thus guided participation is an interpersonal process in which people manage their own and others' roles, and they facilitate or limit access to situations in which they observe and participate in cultural activities. It emphasized routine, tacit communication and arrangements among participants. Therefore, communication and coordination are central to Rogoff's concept of guidance participation. These collective endeavors in turn constitute and transform cultural practices with each successive generation.

The concept of participatory appropriation refers to the process by which individuals transform their understanding of and responsibility for activities through their own participation. The employment of this concept is also an intention to understand learning from involvement in sociocultural activity. It is a personal process whereby, through engaging in an activity, participating in its meaning, gaining facilities, taking responsibilities and making contributions, individuals change and get prepared in subsequent similar activities. From this perspective, learning and development is a process of becoming through guided participation, rather than acquisition.

In the evaluation of learning, sociocultural theorists take account of individuals' roles, changing participation, coordination with others, the nature of the activity and its meaning to the community (Rogoff, 1997). In order to make explicit understanding of learning through these thee planes, Rogoff (1997:280) suggests the following indicators.

(1) the role people play (including leadership and support of others' roles), with fidelity and responsibility;
(2) the changing purposes for being involved, commitment to the endeavor, and trust of unknown aspects of it (including its future);
(3) the flexibility and attitude toward change in involvement (interest in learning versus rejection of new roles or protection of the status quo);
(4) the understanding of the interrelations of different contributions to the endeavor and readiness to switch to complementary roles (e.g. to fill in for others);
(5) the relation of the participants' roles in this activity to those in other activities, with individuals extending to other activities or switching to different modes of involvement as appropriate (such as skillfully generalizing or switching approaches to participation in certain roles at schools and at home, or to involvement in several different ethnic communities);
(6) how the involvements relate to changes in the community's practice (including their flexibilities and vision in revising ongoing community practices).

Drawing its empirical data from children's development, Rogoff's work hasn't been significantly related to adult learning theories, because the nature of the guidance and participation might vary from children to adults. However, in this study, I find her work useful to understand learning in a general way, since the three planes of analysis not only established a frame to see the individual-context relationship, but also importantly, provides a tool to examine the learning process through interpersonal engagements and arrangements as they fit in social cultural processes. As Rogoff (1995) notes, the concept itself is offered as a way of looking at all interpersonal interactions and arrangements. In relation to my research context, Rogoff's proposal of indicators will lend a practical tool for the analysis and evaluation of learning. However, I argue that Rogoff's concepts are in general limited within the community and institution setting without going beyond this level and taking the influence of social and cultural factors into account. In her works I miss the exploration of the nature of knowledge in depth at philosophy level.

3.2.4 Learning as Identity Development

One particular characteristic of this perspective of examining learning from the contemporary constructivist-sociocultural perspective is the recognition of the significance of identity[2] development when examining learning (Du, 2006). The basic belief is that learning is about becoming a person in a society (Jarvis, 1992, 2009), and learning makes people become who they are and creates personal histories of becoming in the context of communities (Wenger, 1998). Therefore, learning is about transforming the experiences of living into knowledge, skills, attitudes and beliefs so that individual human might develop (Jarvis, 2003, 2009). Through participating into communities of practice, people define who they are by interpreting the engaged experiences as well as the interactions with other participants. This learning process is also a process in which people develop their identities (Wenger, 1998, 2009).

Ways of defining identity vary. In this study, I will take the perspective of learning and employ the definition of identity by Wenger (2004:10), that is, 'a learned experience of agency in the context of social structures.' Different from a self-image, which is described in words, an identity entails an integration of experience and its social interpretation, because it involves not only reified categories, a personal trait, a role or a label, but also a lived experience of participation into specific communities (Wenger, 1998).

In this aspect, Wenger shares similar ideas with Jarvis (1992, 2001, 2009) and Barnett (1994), who both have similar notions that learning is more than having knowledge and gain certain competences, it also involves being and becoming a kind of person in the society. In the study of learning, the writers see the two sides of the coin: one side, learners take in the transmitted knowledge and conform to the established practices, beliefs and values; the other side, learners actively participate in the process of creating knowledge and building up new practices, beliefs and values. By asserting the notions of identity, being

[2]In this study, the choice of the term *identity* had a departure from a sociocultural perspective to examine the construction of self. In contrast to the psychologists' biological determined notions of *personality* and *characters*, it highlights the role of social interaction and contexts, and implies both of the conforming and active roles individuals play in identifying oneself and in influencing the social development (Jarvis, 1992).

and becoming in the study of learning, these writers share a common interest in underlining the active, participative, communicative, and reflective parts of learning for individuals.

The perception of learning as identity development has a profound connection with concepts of experiences, practice and participations in a specific social context. People define who they are through experiencing activities. Participants of the activities engage with each other and relate to each other, in this way, they form a community where they develop practices and learning together. Participation is a process in which people become a member of the community. The membership in a community of practice demands the evolution of learning in the practice of the interaction between an experience of meaning and a regime of competence. And this interaction of experience and competence is, in turn, the potential for a transformation for learning individually and collectively in a meaningful way. In this process of participation, people create their personal histories of becoming and change the self. Learning takes place, as learners become full participants of a community by increasingly taking on the roles and responsibilities of community experts.

In brief, this way of perceiving learning through identity development covers not only cognitive part of learning like competence and knowledge, but also affective part of learning like emotions, relationships, social existence and the engagement in the world — the interpretation of these, according to Wenger (1998, 2004), make people develop their identities.

In relation to engineering education, an engineering university represents the professional culture where engineering values and activities are practiced. Thus, the learning culture encompasses academic training for expected competencies, social interaction routines and established practices for doing engineering work. In practice, studying engineering in the PBL study form at the engineering programs at AAU, which is characterized by doing projects in groups, involves not only mastering of technical skills, but also adjusting to attitudes, principles, responsibilities, values and expectations that provide messages in perspective of constructing a professional engineering identity.

Perceiving learning as identity development provides a new perspective to examine both learning and education in a school setting, in particular, since the identity work process of learners has not drawn enough research attention in the development of engineering education (Kolmos, 2006). However, in my study, I argue that in order to employ this approach as an analytical tool, it demands a basic assumption about a sustainable learning context to ensure the entry of participants. As Wenger's work was questioned by Tenant (1997): what would happen to learning in situations where the community of practice is weak or exhibit power relations that seriously inhibit entry and participation? In my study, an extension of the question can be asked: what will happen to the ways people develop their identity when there is unequally shared power?

3.3 Constructivist-sociocultural Approach and Bildung

A review of the above discussed theoretical works provides different possibilities of examining learning under the umbrella of constructivist-sociocultural paradigm. What is agreed by the different constructivist-sociocultural works is that learning takes place from the interaction between the individuals, and it is a changing process in a certain sociocultural context. In addition, in relation to a setting of formal education (for example engineering education), the discipline and learning contents also play an important role. As summarized by Rogers' (2002:88), there are four shared focuses in the contemporary adult learning theories: (1) focus on who is doing the learning; (2) focus on the context; (3) focus on the kind of learning task being undertaken; (4) focus on the processes involved.

A related concern of perceiving learning and education with those focuses can be found in the concept of Bildung,[3] which is well applied in the German and Scandinavian settings (Henriksen, 2006). In educational philosophy, Bildung is a concept that is associated with knowledge and learning, and closely related to pedagogical, social, cultural and political issues (Wimmer, 2003; Koller, 2003; Bauer, 2003; Masschelein and Ricken, 2003; Shaafer, 2003). According to Masschelein and Ricken (2003), the contemporary use of the concept of Bildung covers two opposite poles: it does not only indicates the cultural contents of education, but also means a specific sort of self-development, and it is used to interpret the western form of life. Between these two poles, it also refers to formal competency of acquiring currently required knowledge. It denotes both the processes — the development of the

[3]The concept of Bildung was introduced in 18th century and articulated in the 19th century in German educational and pedagogical discourse. It continues to be a key term in educational thinking, though its meanings are expanded, and it turns to be more problematic and increasingly contested. In this study, this concept is only used for its application in the educational setting (Masschelein and Ricken, 2003).

personality or identity — and the results of learning. It also signifies an integration of social transformation through the formation and cultivation of the individual (Masschelein and Ricken, 2003:140).

The Bildung concept also provides a perspective to understand education. According to Schaafer (2003), education is distinguishing itself by the claim of evoking more than knowledge and capability. Instead of knowledge transferring, it should be a place for individuals to learn and develop the self. Accordingly, educational settings should not only be places for a confident command of knowledge and learning possibilities, but also be places to realize the qualification of self-determination (Bauer, 2003). So education should aim at cultivation of the personality, at competencies which constitute the individual as an enlightened and self-determining subject.

Relating the concept of Bildung to the constructivist-sociocultural perception of learning, some common concerns can be witnessed: (1) focus on the process of learning; (2) consideration of the institutional and societal settings; (3) recognition of the identity construction and transformation process; (4) identification of changes that are involved in the learning and transformation process. Therefore, this study suggests a combination of these two approaches for the following reasons and benefits: (1) to gain more understanding of learning through relating perceptions from different academic traditions in the western world.[4] (2) To gain more understanding of learning in the structure of individual-institution-society, in which learning function as both the goal and outcome of social practices. (3) To gain more understanding of learning in educational practices through examining the role of educational institutions (particularly universities in this study) as a learning organization in the knowledge society. As has been discussed in Chapter 2, higher education in general is facing challenges with respect to its role on learning and education. Relating constructivist-sociocultural learning theories and Bildung concept will lead to recognition that learning processes in educational institutions will cover the formation and transformation process of a self, and with knowledge and ability of cooperating, reflecting and coping with soci-

[4]In this study, the constructivist-sociocultural works I draw on are mainly from English-speaking cultures. The concept of Bildung is well-applied in the German and Scandinavia cultures. In the Danish context, there is also a concept of Dannelse that can be related to the Bildung concept, which I do not have room for discussioning in this study.

ety in general (Henriksen, 2006). The application of Bildung concepts in the contemporary educational system establishes a complex web that represents all the concepts that linked with education, for example, pedagogy, instruction and socialization, and at the same time constitutes the transcendental unity of the meanings and symbolizations of those concepts (Wimmer, 2003). In this way, Bildung can be regarded as both a goal and the process of education and learning in the knowledge society.

3.4 An Analytical Tool of Understanding Learning

A central question in the constructivist-sociocultural approach is the relation between structure and agency. In practice, different theories give primacy to one or the other (Roger, 1996): some to the structure — putting more weights to the analysis of issues of cultures; and some give more significance to agency — underlying issues of individual actions. Another highlight of this approach is the recognition of the importance of context. In practice, each of them provides one important perspective of seeing learning and has different concern in the application to the context. For example, Lave and Wenger, Rogoff, and Wenger extensively address off-school settings as learning context, whereas Jarvis gives considerable concern to the role of school education as milieu for learning.

Thus, a comprehensive method is needed in order to understand learning and its context that cover both agency and structure. As a response, both Jarvis (1987) and Wenger (1998) propose a social theory of learning[5] by providing an intermediary analysis for the 'ongoing, mutual construction of the two' (Wenger, 2004:6), in which learning brings structure and agency in close interaction. This goes in tune with the contemporary application of Bildung concepts, which cover both personal processes and a social process as Bildung takes place in a certain social context (Henriksen, 2006).

[5]Both Jarvis (1987) and Wenger (1998, 2004, 2009) propose a social theory of learning, which have different focuses but share similar departure of intending to bring together social theory and learning theory and benefit both of the theoretical areas. In this study, I place their works into one category with the focus on their shared belief that human learning is fundamentally social, no matter it takes place in social interactions, in a group or by oneself. This theory is distinguished from a theory of social learning, and as Wenger puts that rather than as a replacement of the mechanics of learning, a social theory of learning means to addresses learning at a different level (2004:5).

3.4.1 Understanding Learning from Three Levels

Based on the belief that learning involves a process of identity development, a relation between Bildung and learning can be witnessed. In his elaboration of the employment of Bildung concepts, Hojrup (see Henriksen, 2006) proposes a way to understand the creation and formation process of the self from three levels: a personal level, a life-form level and a societal level. Based on this inspiration as well as the above discussed theories from constructivist-sociocultural approach, a model of understanding learning is developed to examine learning at three levels: individual/personal level, community/institutional level, and sociocultural level. This is an attempt to examine learning through a web that encompass a weave of factors like individuals, different levels of educational institutions, and broader social cultural context as well as the interaction of these factors. In this way, education and learning is related to both individuals and broader social transformation.

Individual/personal level

At this level, rather than searching for the nature of internalization[6] as a medium from external knowledge or skill to an internal depository, the basic understanding of development and learning lies in the investigation of people's actual involvement in activities. Learning involves much more than individuals' possession or acquisition of storage of knowledge. It becomes more effective when learners have the awareness of participation and reflection on experiences (Kolb, 1984). Participation in the sociocultural activities in communities of practice involves much more than mastery of the technical knowledge or skill associated with undertaking some task (Rogoff, 1995; Wenger, 1998, 2009). It leads to a changing process of a person from being relatively peripheral to the center (being experts) (Lave and Wenger, 1991),

[6]These writers distinguish their perspectives of development and learning from the perspective of 'internalization' in two main aspects: (1) sociocultural perspective views development as a dynamic, active; while internalization perspective assumes static entities in the 'acquisition' of concepts, memories, knowledge, and skills and so on; (2) sociocultural perspective focuses the mutual process involved in people's participation in cultural activities, while internalization perspective implies a separation between the person and the social context. (3) From the sociocultural perspective, development is seen as transformation, whereby events are coherently related to each other and changes take place continually without division of time into units of past, present and future; from the internalization perspective, development is seen as accumulation with segmented time as from past, present to future.

whereby learners achieve participatory appropriation (Rogoff, 1995) through observing and carrying out different roles, taking responsibilities for managing activities. Bildung at personal level involves a process self-determination, self-development and self-perfection (Bauer, 2003).

However, learning at this level does not exist alone. Participation is 'guided' (Rogoff, 1995) since it involves personal interaction and communication based on cultural and social values. This is also a process of negotiating new meanings. People are always simultaneously dealing with specific situations, participating in the histories of certain practices and involved in becoming certain persons (Jarvis, 1992). Therefore, learning and being/becoming are intertwined (Wenger, 1998).

Community/institutional level

The focuses at the institutional level are put on the institutional structure and cultural influences on intellectual activities. It encourages the recognition that endeavours involve purposes, resources, and practices that are influenced and constrained by the cultural values defined in community or institutional terms (Wenger, 1998, 2004, 2009; Rogoff, 1995, 1997). Relating Bildung to organizational settings allows seeing self-development in the process of getting involved in learning activities that are related to professions.

In Wenger's (1998) work, the concept of community of practice[7] is developed to understand how structure and agency meet through learning in a context. The community and its practice represent a social structure, while membership and engagement in practice represent agency. For Wenger (1998, 2004, 2009), learning is perceived as a constant and meaningful engagement in shared practices, in which they interact with others and with the context. These practices are thus the property of a kind of community created over time by the sustained pursuit of a shared enterprise. Therefore, communities of practice, according to Wenger (1998, 2004), are groups of people who share a concern or a passion for something they do and who interact regularly to learn how to do it better. They are formed by people who engage in a process of collective learning in a shared domain of human endeavour. This concept is defined from three dimensions: (1) the domain, that is, the mutual engagement

[7]The concept of community of practice is originated from the work of situated learning by Lave and Wenger (1991).

that binds members together into a social entity; (2) the community,[8] where members engage in joint activities and discussions, help each other, and share information; (3) the practice, through which people develop a shared repertoire of resources: experiences, stories, routines, sensibilities, artefacts, vocabulary, styles tools, ways of addressing recurring problems.

This perception might be agreed by Jarvis' (1992, 2001) when he notes that learning takes place in an interaction between the self and others, and in the context of prevailing beliefs and attitudes. For him, learning is 'the process of individuals constructing and transforming experience into knowledge, skills, attitudes, values, beliefs, emotions and the senses' (2001:63). Jarvis also emphasizes learning as social action and interaction rather than social adaptation. The relation between individual and society is more complex than a one-way transaction, that is, individuals' merely passive adaptation and conformation to the prevailing cultural value of the society. The individual-society relation involves interaction and mutual influence, as he notes:

> All aspects of the individual are, to some degree, a reflection of the social structure. But this is not merely an acquisition or receptive process ... individuals actually modify what is received and it is the changed version that is subsequently transmitted to other people in social interaction.
>
> (Javis 1987:14)

The focus on the nature of cultural understanding helps make what is taken for granted visible (Rogoff, 1995). The negotiation of meaning is embedded in the practice of specific communities (Wenger, 1998). These communities and their practices provide materials for learning — language, artifacts, interaction with others and interpretation of the world.

Sociocultural level

At this level, the focus is the individual-society relation. Learning involves being a person, thus the learning process is also a personal growth and development process, which are cultivated both through learning transforms the

[8] According to Wenger (1998), the main character that differ the term 'communities of practice' from other uses of the word 'communities' is that participants have learned through their mutual engagement in these activities.

engagement in the world as well as being in the world (Jarvis, 1992, 2009). Learning is a therefore social becoming, the ongoing negotiation of an identity that people develop in the context of participation in communities and their practices (Wenger, 1998, 2009). Bildung at this level emphasizes the individuals' abilities to function as able and resourceful members of society as citizens who are reflective, collaborative, and responsible (Henriksen, 2006).

The three levels are intertwined and inseparable. Firstly, in the context of communities of practice, participants are involved in a set of relationships over time and communities develop around things that matter to people. In this way the concept of Bildung also offers a way to get knowledge about, and relation towards us and others. As Masschelein and Ricken (2003) note that the question of our being is a question of being-together. People are given a sense of joint enterprise and identity in the process of organizing shared knowledge and activity. Therefore, the formation of a community of practice is the negotiation of identities. Inevitably, the practices deal with the profound issue of how to be a human being in this community. At the same time, community membership gives the formation of identity a fundamentally social character. The interactions involved, and the ability to undertake larger or more complex activities and projects through cooperation, bind people together and help to facilitate relationship and trust.

Secondly, the characteristics of a community of practice are cultivated in the historical, social and cultural context, and the experience of identity in practice is also a way of being in the world. The dynamic process of formation and cultivation of the individual and a social transformation process is specially addressed in the Bildung concept (Masschelein and Ricken, 2003). This integration denotes what educational institutions and educators do as well as designate how individual is the product both of his/her own activity and of that of the society (Schaafer, 2003). Therefore, this model, seeking to link individual activities with broader cultural practices lends a tool to understand both the structure of the contexts for learning and how learning occurs in them.

3.4.2 Meanings Seeking and Identity Development as Core of Learning

Two key elements stand out of the discussion of this model of understanding learning: seeking the meanings and developing an identity in the learning process. Learning first of all is to seek meanings from the participation and engagement in activities (Rogoff, 1995) and the practice of experiences (Wenger, 1998, 2009). The effectiveness of learning lies in, as suggested by Lave and Wenger (1991), engaging in practice, rather than being its object. The purpose of learning is for an individual to construct his or her own meaning, not just memorize the 'right' answers and regurgitate someone else' meaning. This is further elaborated by Wenger when he suggests that meanings arise out of a process of negotiation that combines both participation and reification, thus 'we derive our meanings from 'doing with others' (participation) and from the reified 'objects' of that doing' (1998:135). In this way, learning is also a way to experience our life and the world as meaningful individually or collectively. As Wenger argues, from both theoretical and practical perspectives, a focus on seeking meanings in the study of learning should be as critical and useful as focus on the mechanics of learning, because it is a level at which learning becomes part of the experience of being human. However, it does not suggest that people learn better in groups or in other interactional contexts or that individual learning is somehow inferior or to be avoided. Nor does it deny genetic heritage. To use Wenger's words, it simply claims that our experience of our genetic given is under culturally based interpretation (2004:4). This premise is significantly essential to understand learning when the focus is on meaningfulness and when learning is put within a social context in which the meaning of learning could be negotiated.

Seeking meanings in the learning process is closely linked with the identity development, because meanings are generated from the way people interpret their experiences (Wenger, 1998). The practice of these experiences belongs to the shared communities, which becomes a way to identify people. Based on the premise that people acquire an identity through social interaction, issues of identity are an integral aspect of this learning model. The following main characteristics stand out in the identity development through the learning process. (1) The work of identity is on going and pervasive. This is because identity in practice is not an object, but constant becoming. As Wenger (1998) notes that it arises out of interplay of participation and reification, and it is something people constantly renegotiate in the daily life and communication. (2) The identity development process is also an identity management process. People's identity tend to become more and more complex when each person has multimemberships of different communities and when people are situated in diverse social contexts where the identities are defined. Thus, the process of learning is also a process of identity management, and a part of identity construction involves reconciliation between various communities. (3) Identities are not fixed. As Jarvis (1992, 2009) points out, people first internalize the identity that others ascribe to them — as child, son or daughter, for example. Then they acquire other identities as they progress through life, through participating into different communities of practice. Multimembership remains a fundamental characteristic of identity (Wenger, 2004). (4) Identities and social structures are reflecting each other, since the processes of learning and membership in a community of practice are inseparable. Identity is a display of competence that is formed through practice and reification in the context of communities. Thus, the unit of analysis for identity is both person and community. On one hand, learning involves transformation of identity — in the sense of both objective capacity and subjective awareness, and even readiness (Jarvis, 1992). On the other hand, the ongoing work of identity always has the potential of reframing social structures.

Summary

Learning is a complicated set of processes, and it is difficult to capture all the features in one perspective of definition. In this chapter, I have reviewed a number of theories of learning, which provide different standing points of understanding learning. Thereafter, I proposed a method of analysis combining the concept of 'Bildung' and relevant learning theories in the constructivist-sociocultural approach in the field of learning and education (specifically engineering education).

The development of a model of understanding learning at three levels provides a viewpoint to see the formal structures such as organizations, schools and so on from the perspective of engagement in practice. It also allows seeing how learning occurs through people's interaction in social activities within formal structures. In this way, education and learning is related to a broader social transformation in the process of globalization and in the development towards a knowledge society.

By locating meanings-seeking and identity development as the core of learning, the model developed in this chapter will be used as an analytical tool to explore the understanding of learning from a perspective of how learners interpret their experiences as meaningful, and to examine how they develop and manage different identities through learning. It will be combined with gender theories (see Chapter 4) and developed into an analytical framework (see Chapter 5) for empirical work in this study.

4

Understanding Gender

This chapter discusses the concept of gender and gender relations as well as their relation to knowledge and learning. It starts with an introduction to the choice of a poststructural perspective as the standing point of perceiving gender in this study. This is followed by a discussion of the social construction of the terminology of gender. The employment of the concept of 'doing gender' (West and Zimmerman, 1991) allows understanding gender from the perception of individual characteristics and from the perspective of social interaction where gender works and is being constantly created. Based on the feminist work of Harding (1986), Hirdman (1990), and Gherardi (1995), the social construction of gender relations is discussed from three levels: cultural level, institutional level and individual level. This perception of understanding gender relations is elaborated through the introduction to the metaphor of 'gender contract' (Hirdman, 1990). The last section of this chapter provides a brief review of feminist argument towards the gender-blindness and gender-bias in the prevailing adult learning theories. This is concluded by Harding's (1996) work on 'gendered knowledge system', which argues that the influence of gender relations on knowledge and learning make men's and women's learning opportunities and experience different.

4.1 The Choice of A Feminist Perspective

Feminism can have different faces in the eyes of different people, and feminism is a changing agency itself in the history of human development. For some

Gender and Diversity in a Problem and Project Based Learning Environment, 57–75.

people, the word 'feminism' itself brings negative response, which stands for an image of aggressive women who are fighting for power all the time and regarding every man as oppressor and enemy. For some people, feminism identifies an image of beliefs, values and commitment. In perspective of learning, feminists place gender at the center of analysis, and choose to understand women's learning within a broader social context, which includes attention to the social determinations of gender roles and norms (Flannery and Hayes, 2000). The employment of a feminist approach in this study is an attempt to bring a gender perspective into the examination of learning.

In the western academic world, the past century witnesses a variety of feminist approaches, a high percentage of which place focus on women's experiences (Flannery and Hayes, 2000). In concern with women's knowing and learning, Flannery and Hayes (2000) categorize the multiple feminist theoretical perspectives into three broad frameworks: psychological, structural, and poststructural feminist theories.

(1) Psychological feminist theories, with 'Women's Way of Knowing' (Belenky et al., 1986) as a prominent example, emphasize an understanding of the differences between women and men, and use these constructs as gender-role socialization. The theories can be underlined in a liberal political perspective that seeks to achieve equality for women and men within the existing social order. Their emphasis in education is on achieving equal opportunities for women to learn rather than on developing a critique of the educational and social structures that oppress women. These thoughts contribute to challenge the invisibility and marginalization of women's experience in the knowledge-building process. However, they are criticized for not questioning the nature of the social structure and not discussing women's behavior and experience with respect to structural causes of women's oppression.

(2) Structural feminist theories have focused on understanding the social structures that contribute to women's oppression. They have attempted to explain how patriarchy leads to gender-based oppression and affects women's status and experiences. The goal of this work is to change the social structures rather than individual change. In terms of education, this framework contributes to

locate women's learning in light of social structures and draws the attention to the reproduction of power relationships in settings like classrooms, helping to understand how groups of women may have different experiences. However, it has been challenged by providing limited picture of women's experiences as individuals.

(3) Poststructural feminist theories place emphasis on individual women's response to their unique and particular experiences of oppression in the specific context. This approach contributes to connect individual experience and social structures in new ways. Language and thoughts have been used as the main means of understanding. The attention to individual agency helps to see the possibility of resistance and change. This framework highlights the notion of consistently shifting identity and social structures, helping to recognize the complexity of women's identities and women's differences as well as similarities. Poststructural theories question the notion of a single truth and suggest diverse versions of truth according to people's life experiences (Tisdell, 2000).

However, there exist potential dangers in this approach (Flannery and Hayes, 2000): first, the focus on diversity might make it difficult to come out with common knowledge and unified action; then the concern with language and thought as representations of experience can make it difficult to draw concrete implications for women's everyday lives and learning. Moreover, it will leave questions of what sort of individual and social change is needed to prevail over domination, thus bringing difficulty in putting this issue to political agenda.

In this study I draw on feminist literature that might belong to different approaches but are rather relevant to and make sense for the research context of this study. In general, my study is most closely related to the poststructural framework according to the assumption in that there can be gender differences in learning with respect to learning opportunities, approaches and experiences in specific sociocultural context. Nevertheless, as I claimed in Chapter 1, this assumption does not mean that all men's learning is necessarily different from or opposite to all women's, nor do I suppose that any way of learning is superior or inferior to any other. The aim of employing a gender perspective is to examine the diversity of the people's learning, based on individual patterns under the influence of the prevailing gender relations, with a focus on both men's and women's experiences in the learning process.

4.2 The Concept of Gender

This section discusses the terminology of gender from a social constructive perspective, which considers gender neither connected to biology nor an intellectual construction, but as being created in social, interactional praxis. This perception leads to the choice of the concept of 'doing gender' (West and Zimmerman, 1990) as the basic standing point of understanding gender in this research. In this way, gender is not just a role, but the product of social doings, which involves participating in a complex of socially guided perceptual and interactional activities in the micro-settings of daily practice (West and Zimmerman, 1990).

4.2.1 The Social Construction of Gender

The terminology of gender is a socially-created and academically-used concept, and in some languages, the word 'gender' even can not be distinguished from the word 'sex'. Thus a common way to make a distinction between sex and gender is to understand them from biological and cultural perspectives respectively (West and Zimmerman, 1991). As Anthony Giddens puts:

> 'Sex' refers to the anatomical and physiological differences that define male and female bodies. 'Gender', by contrast, concerns the psychological, social and cultural differences between males and females; it is linked to socially constructed notions of masculinity and femininity.
>
> (2001:107)

In the history of western societies, the accepted cultural perspectives on gender views women and men as naturally and explicitly defined categories of being. The fundamental differences between men and women, that is, gender differences, are supported by reproductive functions, division of labor into men's and women's work and feminine and masculine behavior and attitudes (West and Zimmerman, 1991). The prevailing division of labor between sexes has been argued by feminists as the main reason leading to unequal positions for men and women in economy, politics and other social positions. According to Giddens (2001), as a form of social stratification in almost all societies, gender not only plays a decisive role in the life chances for individuals and social groups but also strongly influences the social role they are expected to play within social institutions from household to public sphere in society. Gender identities in consequence emerge in relation to perceived sex differences in society and in turn help to shape those differences.

Some feminists argue that gender differences continue to serve as the basis for social inequalities in spite of advances women have made around the world. How gender patterns and gender inequalities[1] are created, sustained and transformed in societies have been the concern of investigations of feminist sociologists in the past decades.

The critique towards perceiving men and women as two distinct categories has been one main feminist critique of Western scientific thinking (Hirdman, 1990; Always, 1995; Gherardi, 1995; Salminen-Karlsson, 1999), because the dichotomous way of thinking itself implies asymmetrical power relations between two poles and means hierarchy that promote the domination of men over women. As Always (1995) points out that, the dichotomizing principle leads to looking for single causes or power relations between two roles — one with and one without power — an image of men having all the power and women having no power. Besides, this principle overstresses the differences between groups of different sexes but omits differences among individuals within either group. The dichotomy thinking itself implies a hierarchical ordering of the two categories rather than two different phenomena. To use Hirdman's (1990) words, there is always a norm and an exception. Men are the norm and women or anything associated with the feminine are regularly the exception in mainstream science and theorizing.

Always (1995) suggests the need to both construct and deconstruct the concept 'woman'. On the one hand, there is need to make the idea of 'women' and 'female' visible, make it something to be taken into account, give it power, and eventually to make it equal to 'men'. While on the other hand there is the need to eliminate the common image of 'woman', to demolish the ideas of women's 'special' characteristics, and to deconstruct the dichotomy where 'feminine' is

[1]Gender inequalities are referred to the differences in status, power and prestige enjoyed by women and men in various contexts (Giddens, 2001:137).

one of the categories. The need to keep the balance between construction and deconstruction of the concept 'woman' reaches out to another important area of feminist theory building (Salminen-Karlsson, 1999), which elaborates that being a woman is a social construction. This process of production, reproduction, and definition is also a progression where power influences the gendered social experiences, that is to say, gender is socially and culturally constructed.

Gender is an academically created concept. According to Hirdman (1990) and Salminen-Karlsson (1999), in Swedish official discourse, the concept of 'gender' tends to be used to substitute 'woman'. In pretty much of the gender focused research, gender is more often used to describe special characteristics of women, in spite that there is a growing body of research on the male gender and masculinity. Salminen-Karlsson argues that in parallel to 'woman', 'gender' can be used as a falsely unifying concept even if the intention is not to prescribe unbalanced power distribution between the two parts of humanity. In this sense, gender can be seen as a composition of characteristics constructed and attributed in social praxis.

4.2.2 The Concept of 'Doing Gender'

As an attempt to understand the individual — structure interaction, this study takes a standpoint of West and Zimmerman's (1991) concept of 'doing gender', which gives a possibility of both attaching gender characteristics to individuals and keeps an openness to see gender working and being constantly created in interaction. Challenging the ways of understanding the terms of sex and gender simply from the perspectives of biology and culture respectively, West and Zimmerman (1991) argue that the relationship between biological and cultural processes is far more complex that they were supposed to be. Instead, they suggest a recognition of the analytical independence of sex, sex category, and gender in order to have a deeper understanding of the relationships among these elements and the interactional work involved in 'being' a gendered person in society.

To make their proposal of 'doing gender' understandable, West and Zimmerman (1991) suggest a discern of the concepts of 'sex', 'sex category' and 'gender'. As they note:

> 'Sex' is a determination made through the socially agreed biological criteria for classifying persons as females or

> males... 'Sex category' is a way to categorize people in everyday life based on sex criteria... 'Gender' is the activity of managing situated conduct in light of normative conceptions of attitudes and activities appropriate for one's sex category.

In concern with the relation among these concepts, West and Zimmerman argue that the categorization of members of society into indigenous categories such as 'girl' or 'boy', or 'woman' or 'man', functions in a distinctively social way. Thus, sex and sex category can vary independently, that is, it is possible to claim membership in a sex category even the sex criteria are lacking. Sex categorization and the accomplishment of gender are different from each other, because the former is directly related to be a woman or a man; however, the later involves configurations of behavior that would be seen by others as normative gender behavior. Therefore, gender involves femininity and masculinity. For example, a woman is no doubt female, but whether she is seen as feminine depends on to what extent her behavior fits into the socially and culturally constructed norm and value about femininity.

By proposing the concept of 'doing gender', West and Zimmerman (1991) provides a perspective of understanding gender as a routine accomplishment embedded in everyday interaction. The differences between men and women that are created by gender are 'not natural, essential or biological', and these differences are created in doing, not in thinking. Gender is instead of belonging, but rather something people achieve through social practice. It is not that an individual 'is' or 'has' a certain gender, but he/she 'does' a certain gender. Gender is created in the socialization process of childhood, but also in interactions among grown up men and women. Doing gender means creating differences between girls and boys and women and men, differences that are not natural, essential, or biological. People learn gender by doing and practically all social interactions are doing gender to some degree, though the way of doing it varies in strength and manners.

Doing gender provides a perception of understanding gender in a context of institutions. The sex category/gender relationship links the institutional and interactional levels. In organizations, people learn to do different professions and learn to become different professionalists in communities and organizations, at the same time they also learn to do things as a man or woman in the situated context. To a broader extent, a society is partitioned by essential differences between men and women. Thus people are unavoidably placed into sex-categories, in turn, doing gender is unavoidable, due to the social consequences of sex-category membership.

However, to do gender is not always to live up to normative conceptions of femininity or masculinity (to relate to community of practice theory, community is gendered and also doing gender, social movements such as feminism can provide the ideology to question exiting arrangement at both institutional and social levels. These social movements can also provide social support for individuals to explore alternatives to the prevailing arrangement based on sex category with respect to both interactional and institutional forces.

In summary, 'doing gender' concept provides a perspective to understand gender as social being through interactions in institutional/organizational contexts. This perspective also allows seeing the mutual influence between individuals and social structure, as well as changes as normative characteristics of social interactions. The employment of this concept as a point of departure in my study paves a basic platform for the understanding of the gender relations in the following section.

4.3 The Social Construction of Gender Relations

One of the great achievements of the late last century feminist thought is the discovery that gender relations are fundamentally social and cultural relations (Hirdman, 1990; Gherardi, 1995; Harding, 1996; Giddens, 2001; Flannery and Hayes, 2000). This theoretical achievement has enabled thought to move away from the constraining assumptions of biological determinism. Based on different feminist work (Harding, 1986; Hirdman, 1990; Gherardi, 1995), this section establishes a framework to understand gender relations from three levels: individual, institutional and sociocultural. This framework will be elaborated through the employment of the metaphor of 'gender contract' (Hirdman, 1990).

4.3.1 Perceiving Gender Relations at Three Levels

An understanding of how gender is produced in social situations helps to clarify the interactional scaffolding of social structure. The end of 20th century witnessed a shift in feminist thought from individualist to structural and symbolic concepts of gender relations. Some major shifts are summarized by Hirdman (1990) and Harding (1996). (1) Gender is a socially and culturally constructed product, which also constitutes normative and social-patterned interactions between men and women in society. Thus, in order to understand 'women' (or 'men'), it is important to look at the relationship between women and men. (2) Gender is produced not based on individual choices, but more fundamentally by social structures and their culturally distinctive meanings. (3) gender exists in institutions and organizations; (4) the theoretical discussion of gender conceptualize a power relation. Hence strategies for improving women's situation must always address issues of power imbalance. (5) Instead of being fixed or transcultural, gender relations are dynamic and historically changing.

Jurgen Habermas (1984) notes that the profound social and cultural community values exist in every integrated social system. He believes that this assimilation occur at three different levels in three different reproduction processes: (1) cultural overlay; (2) social integration; (3) socialization. From a feminist perspective, gender relations manifest themselves in diverse ways in different contexts.

American feminist philosopher Sandra Harding (1986) has a widely cited theoretical frame for dividing the aspects of gender into three social spheres. First, gender symbolism, that is, gender dualism — general conceptions of masculinity and femininity in different cultures; second, gender structure, which refers to gendered labour division in different organizations; third, individual gender, that is, the socially constructed individual identity and the principles of their social behaviour. Harding also suggests that though the definition of masculinity and femininity differ from culture to culture, they relate to and interact with each other within any culture (1986:18).

Swedish feminist historian Yvonne Hirdman (1990) has a similar proposal to examine gender relations from three levels: cultural overlay — the intellectual images of masculinity and femininity as a whole in the society; social integration — institutional level, that is, division of labor between sexes in all organizations; individual level — individual identity in the society and gender roles. And the three levels interact with each other consistently.

Similar notions can be found in the work of the Italian feminist scholar Silvia Gherardi (1995), when she suggests an analysis of gender relations through examining social practice of femininity and masculinity in the context of organizational culture. Gender relations are illustrated through social interaction in the community, which is engendered by the symbolism of gender in the social structure. People achieve gender citizenship in the context of gendered community; whereby 'how gender is done and how gender can be done' (1995:4) is defined by the broader social patterns.

Relating the works of these three feminist scholars, an analytical tool of examine gender relations can be established. A weave of these three levels provide a pervasive way of understanding gender relations as individual, organizational and sociocultural phenomena. As Gherardi (1995) elaborates,

> Gender — as a socio-cultural product — is constructed at both the symbolic and the interactional levels. It gives rise to social structures which reflexively institutionalize gender relations. Persons 'do' gender when they celebrate each other's gender identity and repair offences to the symbolic order of gender. As they do so, they defer the meaning of gender as they cope with the ambiguity of the blurred boundaries between universes of meaning.
>
> Organizations do gender too Gender is an organizing principle and an organizational outcome. Gender characteristics are presupposed, imposed on people and exploited for productive ends, and there are organizational dynamics which create them ...
>
> (1995:185–186)

These three spheres, as Harding (1986) puts, support as well as strain each other. In this light, the overall masculinity and femininity supports and is supported by the gendered division of labour and the individuals' gender identity. At the same time, some social factors that will change the division of labour in society might influence both the general and individual conception about gender. Individuals might not always stick to the image and behaviour defined by the common conception. Therefore, gender symbolism is not fixed, but consistently changing little by little in the long process of negotiation between men and women, masculinity and femininity.

4.3.2 Metaphor of Gender Contract

From a historical perspective, Hirdman states that gender relations reflect mutual perceptions, and interpersonal agreements have existed between men and women showing them what social roles they are expected to play and how they can promote interaction. The concept of gender contracts is thus used to refer to these agreements:

> These contracts are drawn up by the party that defines the other and exist between individual men and women as well as in the form of a more societal contract about the limits, contents and forms of the abstract genders ... They explain how the fact that there are always perceptions as to how the relationship between the sexes ought to be.
>
> (Hirdman 1990:16–17)

Salminen-Karlsson gives further interpretation of this concept through her study:

> Gender contract theories imply that the subordination of women occurs in mechanisms that are perpetuated by both women and men. Men and women have 'contracts' on how gender is to be done. Gender contracts are primarily about power relations, disguised in many different kinds of 'regulations'.
>
> (Salminen-Karlsson 1999:37)

Thus, the concept of gender contracts can be understood as the embedded structure of conceptions, expectations and understandings that standardize the interaction between people (Salminen-Karlsson, 1999). The contracts have been drawn up by the male side that has more power, but women are co-creative and co-active creatures — they are just as integrated into the process of defining, imposing and passing down as men, despite their lower social status (Hirdman, 1990:16). In this way, this concept provides a tool to measure the limits of women's opportunities, and also, deepens our understanding of how history has endured the primacy of male norm.

Gherardi (1995) also proposed a similar metaphor — the sexual contract — to explore the pervasiveness, elusiveness and ambiguity of gender in organizational culture. From a discourse perspective, Gherardi describes how daily activities in the organization are engendered and how masculinity and femininity are constructed through humours, jokes, language, dress, manners, according to the sexuality-based social expectation. The gendered features of organizational activities implicitly select people based on the socially constructed gender features, however, the interaction between the gendered activities and gendered people is 'inscribed in the social imaginary that projects the cultural elaboration of gender on to nature and then imports it back into the culture as natural difference' (1995:40).

According to Hirdman (1990), gender contracts seem like a self-sustaining system with both men and women participating in the continuous process of defining and reinforcing social behavior and expectations. However, Hirdman claims that it is also an ongoing process of negotiation and transformation. The weak party tries to negotiate and make small changes in the contract for their benefits. For example, when an increasing number of women got jobs in industry or companies, the gender contract started to dissolve. Industrialization, technological development changed the limits and destroyed previously established routines concerning marriage, sexuality, etc. Thereafter, the rules of the gender system were revealed, questioned, defended and reconstructed. The system is a process of continuous defining and reinforcing of behavior and expectations. Hirdman believes that the less the rules and practices of separation between the sexes operate, the more the primacy of the male norm will be questioned and the more illegitimate the male position will be.

Therefore, Hirdman (1990) believes that there is room for redefinitions by certain social changes and there exists the power to change the system with the human thoughts. The more the intellectual components dominate over the biological, the greater potential there is for change. Some possibilities for making changes of the gender contract are suggested by Hirdman:

(1) To reconstruct gender contracts by challenging the basic dichotomies in human thinking and doing. This can be done by breaking the rules of separation of the sexes — letting women enter men's area and to perform men's tasks. With more women entering

men's world, and taking men's tasks, it will be easier to overstep the boundary and transform the male norm in society. This idea is shared by Gherardi when she puts that 'if we are to ensure that social differentiation is no longer based on sexual differentiation, we must destabilize all thought which dichotomizes (either male or female) and hierarchies (male as One, the norm; and female as the Other, the second sex)' (1995:4).

(2) To violate the gender contract by confronting the primacy of male norm, when men are reduced to the same powerless role as women. This can take place in societies where power is moved out of human beings to technology, and when men and women are pushed together into a common humanity.

(3) To change of gender contracts through the influence of other changes in the society. As discussed by Sundin (1997) and Salminen-Karlsson (1999), changes in technical development, economic development or organizational changes may also lead to re-negotiation and changes of gender contracts.

Salminen-Karlsson (1999) discuss about the inherent tensions and different kinds of contradictions among the three levels of gender relations. Hence, there can be contradictions between the different levels (for example, between the societal equality ideology and the gendered division of labor or ideology of individuals at an institution), or between different social contexts (between educational institutions and the labor market). These contradictions may open a change for re-negotiation of gender contract.

Therefore, gender contracts, defining tasks, locations, characteristics, etc. of the two genders, have changed slowly or rapidly at different stages of human developing history, and they are changing constantly. However, my study finds some untouched questions in the discussion on changes. (1) the proposal of making changes in gender contract by forcing or inviting more women into men's sphere and perform men's tasks does not include explicit indicators of the change of male norm. (2) the suggestion on reducing men's power remains theoretical assumption without empirical support. (3) I argue that Hirdman places focuses on the discussion at a structural level; however, there is a lack of interest in the experiences of individuals. These questions will be examined in my study through the empirical work.

4.4 Feminism Confronts Learning Theories — A Historical Review

Since late 20th century with the development of social, cultural, economical, political factors, the social norms and gendered roles are undergoing various changes in the western world. This leads to a remarkable growth of women's participation in both formal educational programs and informal learning activities. Many efforts have been made in different western countries aiming at improving learning opportunities for women. However, these practices did not lead to significant understanding of women's learning at a philosophical level (Flannery and Hayes, 2000).

Since 1980s, the prevailing learning theories in western academic world have been challenged by feminist scholars for the lack of consideration of gender in the broad philosophical standpoint in adult learning theory (Gilligan, 1982; Flannery and Hayes, 2000). Different feminist works arose — questioning the general negligence of women's experiences along the historical development of psychology. These studies are regarded as representatives of an important phase in feminist theorizing (Maher, 1996).

Gilligan's work (1982) has been regarded as a classical representative of feminist work during this period. Not only showing concern with women by reporting how women's selves are formed through involvement in various learning processes — especially informal activities — in particular social contexts. Her work also established a standing point to examine the creation of knowledge by analysing how knowledge is generated through these different forms learning in women's experiences.

In her work, Gilligan (1982) described women's stories about their experiences as well as perceptions on marriage, family and work relationships and found that there were profound differences between the development of males and females, which originated from different experiences of parenting. For example, women tend to find their identity more within relationships with others. Based on different literature on women's development, Gilligan summarizes some features that are unique to women — their embeddedness in lives, their orientation to interdependence, their subordination of achievement to care, and their conflicts over competitive success leave them personally at risk in mid-life. However, Gilligan argues that this observation reveals the limitation in an account which measures women's development against a male standard and ignores the possibility of a different truth. It is 'more a commentary on the society than a problem in women's development' (1982: 171). Based on this, she criticized that much of the previous literature about human learning has drawn conclusions about people in general from research based only on males, especially in educational research. Gilligan's work made a significant contribution to the developmental understanding of human knowing with rich empirical data. It also worked as a turning point in the research on knowing by bringing a new perspective — the inclusion of women's experiences, as she notes 'women's morality can throw new light on our understanding of morality more generally.'

Inspired by Gilligan's work, some other American feminist research Belenky, Clinchy, Goldberger and Tarule investigated the development of 'women's ways of knowing' (WWK) (Belenky et al., 1986). They interviewed in detail 135 women with different backgrounds — asking them questions about self-image, relationships of importance, education and learning, decision-making in life, moral dilemmas, accounts of personal changes and growth, visions of future, etc. (1986:11). They confirmed and extended Gilligan's argument that women have a more relational orientation, which lead to different perspective from men. Through looking into individual women's experiences in terms of personal development from knowing, they categorized five ways of knowing. From these forms of learning, WWK study provides a typology of how women learn and how knowledge is generated through women's learning.

In WWK (1986), focus is placed on the genderization of knowledge and knowing based on women's development in a patriarchal society. Their stories depicted a variety of different ways women understand, accommodate, and resist societal definitions of authority and truth. By studying women's development and education, they did not just intend to describe different ways women know, but how women (in the U.S.) are socialized to know and how they respond to socializing forces. The authors contribute to the growing understanding of how gender is constructed in the lives of diverse women, although this work has been challenged as putting overweight on single category of women due to the absence of consideration on race, class and culture (Harding, 1996). The work of 'WWK' plays an inspiring and influential role in late works on women's ways of knowing and learning. For example, in relation to higher education, 'WWK', combined with other feminist writings, created considerable discussion among educators and inspired some women-friendly instructional approaches as well as curricula reforms (Goldberger, 1996).

In brief, as Flannery and Hayes (2000) summarize, the end of the 20th century witness a status of feminism on learning, in which feminist scholars started to challenge the prevailing learning theories for their significant biases toward certain values and cultural norms, which are mainly based on men's experiences and the assumption that women's experiences are consistent with men's. The undertaken analysis of gender in learning has simply relied on identifying consistent patterns of differences between women and men as two categories rather than providing more nuanced insights into diversity among individuals in either gendered group.

4.5 Gendered Knowledge System

Based on a review of feminist works on gender relations and feminist arguments on gendered ways of knowing, Harding (1996) writes about the effects of gender, as a system of social relationships, on knowledge and learning. Focusing on the role that power and culture play in knowledge construction and strategies of knowing, she argues that gender can be treated as distinctive cultures, as the term she uses — 'gender cultures'. In many social situations, men and women occupy the same cultures, that is to say, many social, classical, or religious cultures contain both genders. However, there are also some single-gender cultures within different other cultures, which means that primarily women or men are to be found in these cultures. It can also be that the subcultures are gender-coded so that the gender of the culture does not necessarily reflect upon the proportion of women and men in it. For example, we use 'masculine' when we talk about cultures of military or sports, even when women are sometimes are visible in them. 'Feminine' is the code of the culture of the fashion world or elementary schools, even though plenty of men can be found in them.

According to Harding (1996), gendered ways of knowing can be witnessed in gendered cultures. Firstly, women and men are confronted with different elements of nature's regularities for both biological and social reasons. People have different physical resources based on different physiological characteristics, for example, men and women might have different preference in sports because of biological differences.

Secondly, when gendered social structures assign women and men to different activities, they will tend to interact with different parts of nature — from the gendered socialization and stereotype perspective, women may have more opportunities to interact with babies and men may have more opportunities to interact with car motors, for example.

Thirdly, the system of gender relations can give women and men different interests and concerns even when they are in similar situations, so the knowledge they have about similar situations may be different. Consequently, their knowledge of the local environment can be different.

Fourthly, men and women can have different relations to the cultural discursive traditions that direct their practices and give them meaning. The system of gender relations may also lead women and men to develop different ways of creating and sharing knowledge. Harding provides examples of how women scientists seem to use skills, resources, and forms of interaction that are different from those used by men.

This can be related to women's learning experiences in engineering. Women have been brought into engineering area in all kinds of recruitment movements, but they have different learning experiences from men, since the learning culture in engineering education has been labeled as masculine. Hence, women might obtain different sort of knowledge and abilities from men even if they are in the same situations. At the same time, women and men might create and share knowledge in different ways within the same institution.

Lastly, Women and men can be found in both cultures, but these cultures shape women's and men's experiences in different ways, thus give them different opportunity to acquire different sorts of knowledge and abilities. Besides, men and women often have different, socially developed ways of organizing the production of knowledge. Again, Harding provides suggestions from research on women scientists, for instance, they tend to seek different research topics, from male colleagues, who normally choose the most competitive new field. Women scientists tend to organize their research teams more around cooperation and less around competitive relations. Women scientists' peer networks tend to have different kinds of resources than those used by their male counterparts. These gendered knowledge systems, like gender relations, as Harding (1996) states, may differ by society, culture, ethnic group, locality and so on, and so may produce different knowledge systems within the cohort of all women as well as between women and men.

Gender knowledge system is closely linked with power relations, as argued by Harding (1996) since all knowledge claims are socially situated, historically local and shaped by culturally distinctive locations in nature, interest, discursive resources, and ways of organizing the production of knowledge. The understanding of gender in perspective of a knowledge system provides another standpoint to look at power relation, especially with respect to the relation between learning, knowledge and power. Based on this understanding, Harding (1996) proposes an ideal situation — by taking ways of knowing of both genders into consideration, that is through legitimating and exploring diversely socially situated knowledge, there is a good chance to expand knowledge while also advancing recognition of the richness and diversity of cultural traditions.

Summary

This chapter discusses the concept of gender and gender relations as well as their association with knowing and learning. From a standing point of post-structural approach, the concepts of 'doing gender' (West and Zimmerman, 1991) and 'gender knowledge system' (Harding 1996) are employed as the analytical tools to explore the understanding of gender, gender relations and their influence on knowledge creation and learning. From this perspective, gender is considered neither connected to biology nor an intellectual construction, but as being created in social, interactional praxis. The prevailing gender relations play an influential role on the way knowledge is constructed and the way men and women take in and participate in the knowledge generation.

This standing point leads to the choice of gender (with focus on both men's and women's experiences) as an important angle of examining learning in this study. Focus on gender does not necessarily suggest gender as the only measurement in understanding human learning, but rather, make a proposal to place value on the diversity of learning based on factors of gender, personal history, cultural background, situated circumstances and so on. The choice of focusing on both men's and women's learning experiences is not based and will not lead to the belief of generalizations about all women and all men. On one hand, there is no intention to show preferences to a particular gendered group and marginalization to the other, on the other hand, there is the awareness that understanding the significance of gender in relation to learning is more complex than assuming women are like this, and men are like that. To use the metaphor from Flannery and Hayes (2000), there is a kaleidoscope of ways of learning, which overlap at times but are unique to different individuals.

5

Gendered Ways of Learning in Engineering Education

This chapter relates learning theories from the constructivist-sociocultural approach to gender theories from a poststructural perspective, based on the belief that an interweave of these two theories will be beneficial to both. This attempt aims to build up a platform to relate these two theories to each other both as a theoretical exploration and as an analytical tool to examine learning and gender relations in the research context. This analytical framework is elaborated through discussing an integration of the three areas of learning, gender and engineering education as well as the intertwined individual-community-society relationships. A review of relevant literature provides background knowledge regarding what has been known in these areas, which sets up a scene as a reference for the exploration of what can be known in my study. This chapter concludes with a discussion of power and change from the perspectives of both learning and gender relations. Two aspects of power are suggested as a tool to examine learning and gender relations as both an individual activity and organizational and social practices. Change as a main characteristic of the agency-structure relationship in the examination of both learning and gender relations involves mutual influencing and transforming processes of both individuals and social structures.

5.1 Relating Learning Theories and Gender Theories

Relating the discussions of relevant learning theories and gender theories in the previous chapters, some shared concern can be identified: (1) the focus on the

Gender and Diversity in a Problem and Project Based Learning Environment, 77–98.

interpersonal relation as well as interaction between structure and individuals social interaction on the understanding of human behaviour; (2) the emphasis of social context in which human behaviour take place; (3) the formation of identities through practices of the interactional activities; (4) the recognition of change as one major characteristic of social phenomena.

However, as many feminist works have challenged the prevailing adult learning theories for their general gender-blindness and gender-bias, this study also argues that gender is not addressed in any of the works of learning that has been discussed in Chapter 3. From a feminist perspective, this study argues that learners' opportunities for participation in activities, engagement in experiences, negotiation for the meanings of practice, and patterns of interaction with others are 'guided' (Rogoff, 1995) by certain prevailing values and norms, among which, gender relations are an important one.

Based on this, this study argues that relating theories from these two areas can benefit each other conceptually and empirically. For learning theories, the introduction of a gender perspective will enrich the perceptions of examining learning and thus increase the diversity consideration. For poststructural feminist theories, the association between learning and gender is innate since doing gender as an organizational practice is an interactional learning activity, in which people learn how to do gender in an appropriate way (West and Zimmerman, 1990), and in which people take in and generate knowledge in

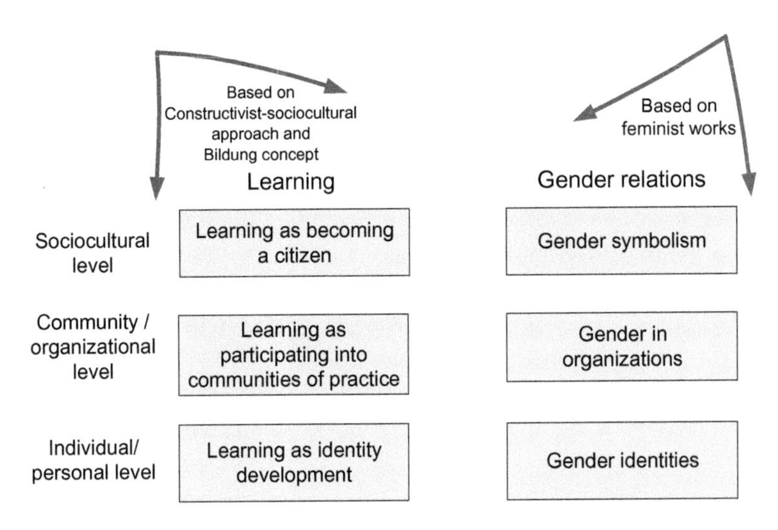

Fig. 5.1 Relating gender theories to learning theories.

a gendered way (Harding, 1996). Relating the two approaches to each other will make a contribution to the exploration of learning and creating knowledge from the perspectives of individual cultivation, organizational development and social transformation.

Taking a feminist point of view, the prevailing gender relations play a role in learning both at individual level and in the social context where learning takes place. Through participating into communities of practice, people know what they know, act according to the shared practice, and learn to become who they are. From a feminist perspective, gender as a major social, historical and political category affects the life choices of women in all communities and cultures (though in different ways in different cultures) (Harding, 1996). From an organizational perspective, gender plays an influential role on social practices, as Gherardi (1995) notes that there is always a gender relation influencing organizational behaviour. People know whether they are men or women through collective activities, and they do gender in an appropriate way and behave as a man or woman through interaction with others, based on both the cultural and local expectation for femininity and masculinity. Accordingly, organizational learning involves learning a certain gender order and a certain way of doing gender (Salminen-Karlsson, 2006). Based on this, this study argues that linking the two areas of gender and learning allows examining the self-development and social transformation through learning processes in a gendered way. Therefore, bildung is gendered; communities of practice are doing gender in organizational contexts, which leads to the gendered experiences of individuals with respect to participation in activities, seeking meanings, and developing identities.

5.2 Examining Gender and Learning in Engineering Education

In this section, the integration of theories of these two areas will be elaborated and related to the research context — engineering education, as illustrated by Figure 5.2.

This figure of gender, learning and engineering education will be elaborated through the discussion of the intertwined relationships among the three conceptual areas and among the three levels. These relations will also be exemplified by relevant research and literature. This aims to make the background

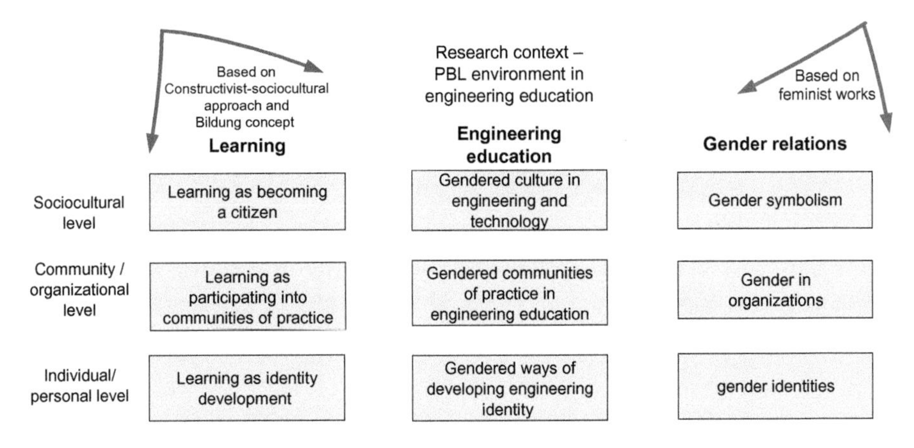

Fig. 5.2 An analytical framework — gender, learning and engineering education.

of establishing this analytical framework understandable before it is related to the empirical work in this study.

5.2.1 The Sociocultural Level — Gendered Bildung

A sociocultural perspective allows understanding both learning and gender as social phenomena. According to both Jarvis (1992) and Wenger (1998), learning involves being/becoming a certain kind of person in the world. From a feminist perspective, one important aspect of being a person in the world involves learning to become a man or woman based on the general conception of masculinity and femininity in that social culture (Harding, 1986; West and Zimmerman, 1990). Relating to the individual Bildung at the societal level discussed by Hojrup (see Henriksen, 2006), learning as becoming is a gendered process. This means that the process of the self-development involves learning to become a gendered person, and to obtain 'gender citizenship' as a member of the society and occupational organizations (Gherardi, 1995). Relating to engineering and technology, a brief literature review indicates that the historical development under the impact of gender symbolism in general defines it a gendered sphere.

Gender symbolism and learning

Gender symbolism plays a significant role in the process of both individual development and social transformation (Harding, 1986, 1996; Gherardi,

1995). Therefore, it is necessary to understand the social construction of masculinity and femininity in the western world. Through the notion of 'hegemonic masculinity', Connell (1987), distinguishes different levels of masculinity in society. By this notion, he refers to the dominant image of masculinity, which, as a cultural norm in western cultures, is in general associated with 'rational', 'independent', 'risk-taking', and 'aggressive' men (This image can be linked with many aspects and need not necessarily match the actual practices or personalities of the majority of men) (Connell, 1995). In her description of the symbolism of masculinity in western culture in general, Gherardi (1995:34) writes about similar vocabularies: 'rational', 'public', 'activity', 'separation', 'thought', 'the mind', 'hardness', 'coldness'.

Hegemonic masculinity is related to feminist debates on 'difference-sameness' (Kvande, 1999), that is, when men are established as the norm of the community, women will be considered as different from, or the same with men, but not the other way around. According to Harding (1986), femininity and masculinity might have different meanings in different cultures, however, no matter in what culture, they always refers to characteristics oppose to each other. This is agreed by Gherardi (1995:34), who used the vocabularies of 'emotional', 'private, 'receptivity', 'connection', 'sentiment', 'the body', 'softness', and 'warmth' to describe gender femininity in western world.

The general gender symbolism in western country defines organizations and professions gendered sphere in the society. For example, engineering, with a close link with the control and mastery over technology (Wajcman, 1991), which goes in tune with the general ideology of masculinity, is culturally defined as male profession and male sphere in many western counties. Accordingly, based on the assumption of 'difference-sameness' (Kvande, 1999), when masculinity is considered as competence for technology (Wajcman, 1991) and femininity as an opposition should be less technological competent than men. Once dealing with hardcore engineering in a career-oriented profession, female engineers are not regarded feminine any more according to the traditional social ideology of women. The following text will provide more elaboration of gendered culture in engineering and technology.

Gendered culture in engineering and technology

In his social theory of learning, Wenger (2004) use the terminology of 'scale' to multiple levels in communities — from local groups to large umbrella communities. All these different levels of communities are nested in a 'fractal' structure — the small and the big scale communities coexist and constitute each other. In this sense, the world of technology and engineering builds up a big scale community in the world (in this study, I mainly refers to western world) and smaller scales of communities locally.

Feminist scholarships on technology and engineering have grown steadily from the early 1980s.

The masculine culture of technology and engineering has been an established fact. Considerable evidence has been found regarding male-norm in the overall engineering community, thus explaining women's underrepresentation in scientific and technological institutions (engineering universities in this study).

Harding exemplifies the influence of gender symbolism on professions through her empirical work on gender relations in the community of science. The discrimination and exclusion of women in science is restricted by the organizational culture of social labor divisions and larger science culture as male norm — if science is regarded as activities appropriate for men, it is not surprising that women are unwilling to develop the skills that are considered as important or successful in these areas.

Harding also criticizes the equity recommendations, which ask women to exchange major aspects of their gender identity for the masculine version instead of questioning the masculine culture of science itself. Because the fact is that, though women have had access to science education and work after long struggles, they are after all 'others' and 'limited to positions just inside the entryway' in the combination of vertical and horizontal segregation (1986:62-63).

Parallel to this, some feminists began to question the patriarchal values of technology itself in the Western culture instead of asking how women could be more equitably treated in a neutral technology. Indication can be found from their empirical work that technology can be classified as one of the strongest symbolic signal of masculinity (Cockburn, 1985; Hacker, 1989; Wajcman, 1991, 2000; Mellstrom, 1995; Berner and Mellstrom, 1997; Sundin, 1997; Salminen-Karlsson, 1997, 1999).

Judy Wajcman (1991, 2000), one of the most influential feminists writing in this vein reviews feminist critiques on technology and confirms the importance of cultural factors in understanding the masculinity of technology — historical and cultural construction of gender leads to the masculine culture of technology and the marginalization and exclusion of women from technology (1991). This conclusion mirrors Harding's (1986) statement about science, and can also be related to the three levels of the interacting relationship in the gender contracts.

Wajcman also points out that technology is political; there are elements of economic and social class interests lying behind the process of production, which stands for power of certain social groups. Here, Wajcman shares these ideas with Cockburn (1985) and Sundin (1997): male dominance is not based merely on numbers, it is symbolic; every man is backed up by men's organization, wealth and ideology in society as a whole, since industrial, commercial, military technologies, and so on, are masculine in a very historical and material sense. Technology is always embedded in its social context and demonstrates men's power of construction and implementing technological change (Sundin, 1997).

In the perspective of sex segregations, Wajcman (1991) agrees with Cockburn (1985): the main point of the value of machined-related skills and physical strength is to make technology masculine, and to make it an important source for men to get power through the monopoly of technology. Technology is thus identified as masculine and masculinity is defined in terms of technical competence. Men's affinity with technology is integral to the culture of technology and the constitution of male gender identity (Wajcman, 2000). It is difficult for women, who have been restricted to the private sphere of household, to have access to the required skills and competence in technological work for innovation. So, the exclusion of women from science and technology is not simply the problem of their lack of physical strength, technical skills and inventiveness, but because of their opposite characters of gender identity.

Engineering is a good example in illustrating the masculine culture of technology in practice. As a technology-based and career-oriented profession, it is hostile to women and contains the smallest proportion of females, since the ideology is constructed that men are powerful and competent and women are inferior and unsuitable for technology pursuits. Cockburn argues against the equal policy opinions that equality can be achieved by recruiting more women into engineering, because there is the fact that women do not want to enter engineering profession themselves. But their reluctance 'to enter' has to do with the sex-stereotyped definition of technology as an activity appropriate for men. If the culture of engineering is masculinity, it is naturally incompatible with femininity. Therefore, to enter this world, to learn its language, women have first to forsake their femininity (Wajcman, 1991:19). This also resembles Harding's (1986) conclusion on preconditions for women to enter the world of science.

5.2.2 The Institutional Level — Gendered Communities of Practice

Organizations, in spite of their claim to be neuter and neutral, are structured according to the symbolism of gender (Gherardi, 1995), therefore, communities of practice are doing gender, which defines the organizational culture as gendered. Gender relations in communities of practice, as tacit knowledge, constrain and shape men's and women's ways of learning differently in terms of opportunities, experiences and the meanings generated from learning. Following Harding's (1996) gender knowledge system, men and women might obtain and create different sorts of knowledge when participating in a community of practice where the culture is defined by the opposite gender. This can be exemplified through the gendered practice of learning in engineering education, where women's participation has been historically hindered and marginalized due to the historical male hegemony as norm.

Gender and learning in organizations

From the constructivist-sociocultural perspective, learning involves both individual behaviour and organizational context. As to learning in organizations, a common view is that individuals take in and process knowledge as well as competence in the context of organization (Jarvis, 2001). As Wenger said, communities of practice are the prime context in which we can work out common sense though mutual engagement. From a feminist perspective, this study argues that it can also be a context where gender contract systems are sustained and the prevailing gender relations are produced and reproduced. Gender relations, as tacit knowledge, might be something we take for granted and tend to fade into the background (Gherardi, 1995). In organizations, the ways people work together, talk, joke, develop beliefs and attitudes is a reflection of gender relations, which thus build up a gendered culture (Harding, 1996), for example, engineering workplaces in western societies have long been established as a male culture; whereas the culture the profession of nursing remains feminine.

These gendered cultures shape the organizational practice as gendered — engineering professional as a male community of practice and nursing profession as a female community of practice, for instance.

Different feminist scholars write about how gendered culture in communities of practice shape men's and women's learning differently. Harding, in her earlier work (1986, 1989) has pointed out that the masculine culture in science institutions have excluded women's involvement and marginalized women participants as 'guests'. In the later work, Harding (1996) internationalizes the focus on culture and draw upon the notion of 'local knowledge systems' to emphasize the necessity of looking at the context. She identifies how the criteria for knowledge acquisition and validation in communalities of practice in the western world hinder and neglect women's contribution to knowledge. In her study about gender in engineering organizations, Kvande (1999) finds that the organizational context and the environment of the community of practice at workplace are created differently for women from those for men, which puts women engineers into conflicts of achieving professional identity and female identity.

Gherardi (1995) also witnesses different learning trajectories and identity development of men and women in the communities of practice in her research. As she finds, when women are situated in male-dominated communities of practice, the positions of women as newcomers in the periphery can be different from those for men — it can be the position of the newcomer on their way to the centre, or it can be a permanent position of marginality. This, to a great extent, depends on whether and to what extent the masculine culture is friendly or hostile to women.

Hayes (2000) discusses the social contexts of women's learning with particular attention to how gender confines women's opportunities for learning. According to her, in formal education, different social settings — from formal teaching materials to tacit knowledge — can influence people's learning, from which people learn important lessons about themselves and relationships with others. These influences can be different for men and for women, because the social settings also play a role in reinforcing gender-based stereotypes. Therefore, to understand women's ways of learning, it is necessary to understand how they respond to the learning resources they encounter in relation to gender.

Hayes (2000) describes three influential aspects of formal educational contexts: curricula, interpersonal interactions, and institutional culture, which can be used as indicators to access the learning context from a gender perspective.

(1) Curricula, which include textbooks, instructional materials, teachers' lectures, and so on, are not simply sources of knowledge about subjects; they are also influential resources of learning about gender. They can strengthen stereotypical roles and images of women and men.
(2) Interpersonal interaction in the educational institution plays an essential role in influencing women's learning experiences as well. Hayes also refers to the gender-based biases in student-student, teacher-student and faculty-student interactions and gender differences in students' response to different relationships. These writers also share the same observation that power relationships are produced and reproduced in the classroom interactions through intentional as well as unconscious behavior of both teachers and students in both genders.
(3) The culture of an educational institution will have both manifest and covert influences on a woman's learning. By institutional culture, Hayes refers to factors like gendered proportion at institutions, schedules of study programs, and intellectual culture, etc. The proportion of women students and teachers in an educational program remains one of the most vital factors deciding the culture. It plays an essential role in reflecting the extent that other institutional factors support women's participation and learning. According to Hayes, the proportion of women students itself can contribute to women's feelings of belong as learners, and to their overall comfort with the learning environment. Women's learning is shaped to some extent by their access to formal education. In the past, women were kept outside of many types of formal education, but women's participation into many educational settings has growing rapidly in recent time. However, the high number of women teachers and learners does not automatically guarantee a supportive culture for women's learning, because, in some formal educational settings, special values are placed on certain kinds of knowledge and ways of

knowing. Thus, the intellectual culture, that is, the types of subjects and knowledge gained can be significantly different for women and for men even though they are present in the same institution of a certain kind of formal education as men.

Gendered communities of practice in engineering education

Relating engineering profession to the life form level of Bildung suggested by Hojrup, Henriksen (see Henriksen, 2006) interpret engineering education as the Bildung of the engineer, which involves both the process (learning to become an engineer) and the product (working as an engineer). In this way the Bildung process for engineering students is not just to learn the application of science to practical tasks, but also a process to be trained engineering competences and to be taught the engineering traditions. From a feminist perspective, the engineering tradition and demands for becoming a member of engineering community is historically based on male norm and closely related to male gender role, which led to exclusion of women as a social group.

An engineering university is an organization where future engineering professionals are cultivated with a combination of both scientific knowledge and technical skills. It can be perceived as the moderate scale of community of technology and engineering, where people learn to become engineers through negotiation of reification (technical knowledge, for example) and participation into the practices in the local context. It can also be regarded as an institution in the organizational level of gender contracts, where gender stereotype can be witnessed from the sex-segregation and the male-norm. Diverse studies have been conducted on how the engineer's occupational identities are created in gendered forms through everyday practices, institutional arrangements and symbolic representations — all the three levels of gender contracts — in micro settings.

Hacker (1989) traces the origin of engineering education from the need of military in history. Military engineering schools shaped all engineering schools in the U.S. in the 19th century. The ideology was to turn out effective, rational, and well-disciplined engineers (mainly male). The military plays an important role in the hierarchies among men and subordination of women historically, which is also a decisive factor leading to the masculine culture of engineering education now.

Berner and Mellstrom's (1997) historical research on the engineering culture in KTH[1] from the end of 19th century also problematizes men's seemingly 'natural' place in engineering education. Through examining curriculum, pedagogical methods and the homosocial life, she found that engineering was not simply technical skills in this technological school; it demanded diligence, capability in problem-solving, and mobility in work; in addition, it involved mastery of power and management with the help of knowledge. Consequently, it meant the exclusion of women. Mellstrom's (1995) study, which is focused on the personal and interactive symbolic forms in understanding the engineering world by both male and female engineers, illustrates that though women have got the right to enter engineering in modern society, the dichotomization of public and private spheres for men and women in this career-oriented world still exists through time and space. The continuity of masculine culture, organization of pedagogy and the homosocial environment still keeps women as 'guests' since 'marketplace masculinity' is the norm for all engineers.

A number of studies have looked at engineering culture in the contemporary western societies. Proposing a synthesis of both gender-role perspective and structural perspective, McIlwee and Robinson (1992) examined women's paths to the engineering world and their college experiences in the masculine culture in the U.S. They argue that the process for women to enter the engineering world is different from that of men, which reflects a peculiar combination of traditional gender roles, academic skills, and a changing structure of opportunity. Women's initial choice of engineering is a function of their skills in math and their practical orientations. But their academic success in the university life is not accompanied by high levels of technical self-confidence because of their lack of mechanical background, their unfamiliarity with technology and their traditional gender role. In their view, the occupational culture and masculine environment in engineering have combined to influence women engineering students' self-confidence in a pessimistic way. A later research by Shull and Weiner (2002) confirms this point and further it by arguing for innovative changes in the teaching methods to foster women's self-confidence while still mastering the discipline.

In relation to McIlwee and Robinson's (1992) study, Dryburgh (1999) describes women students' experiences in Canadian engineering education,

[1]KTH: Royal Institute of Technology, a Swedish technical university.

and argues that engineering school experiences for female students is also the process of adjusting to the masculine culture as well as to the technology-based profession. Women have to work harder than men, and they face obstacles not there for men of similar ability. To conclude, engineering is a more difficult task for women than men, given engineering's masculine identity.

Classroom research looks closer at the male norm in the engineering culture. Hacker's (1989) ethnographic research witnesses the link between technology and masculinity from classroom jokes about technical elitism, private languages, metaphors, childhood pleasures, and adult pursuits at MIT.[2] As she notes (1989:49), women's entry brings contradictions to the engineering world as well as to themselves and as a cost, they give up themselves and become 'one of the guys'. Similar findings about mild profanity, semi-sexual humors, and symbolic violence metaphors can be mirrored in Tonso's study (1996), which suggests how the cultural norms of engineering education works to the disadvantages of women students and faculty even in a nontraditional classroom. For the women who have been accepted here, they must learn to accept existing norms and not openly resist or challenge them.

Besides, both Tonso's (1996) and Salminen–Karlsson's (1999) observation and interviews with teaching staff in engineering universities suggest that the gender-blindness or even male dominant mind of male teachers, who take up the majority of teaching staff, contribute a lot to the construction and reproduction of masculine customs.

The above-discussed research examined how the culture of engineering education at the mediate scale of communities meets the anticipation of males and leads to women's under representation and invisibility through daily practices.

5.2.3 The Individual Level — Gendered Identity Development

From a constructivist-sociocultural perspective, learning involves an identity development through the process of interaction in a social context. From a feminist perspective, through social interaction, people also try to do gender in an appropriate way based on the cultural value of masculinity and femininity. Therefore, the identity development is closely linked with and influenced by

[2]MIT: Massachusetts Institute of Technology in the U.S.

the prevailing gender relations. People can obtain multimembership when they participate in different communities of practice at the same time. Thus, learning also involves an identity management process. However, conflicts might arise among different identities, for example, when women enter the male-dominated profession like engineering, their gender identity is confronted by the culturally defined identity of the community, which is associated with masculinity. Thus, women engineers are facing dilemma situation in their identity management process.

Gendered experiences of managing identities

When Wenger (1998) notes that the experience of identity is the practice of being in the world, there is an indication of a profound connection between identity, experience, practice and the social context. An identity entails an integration of experiences of participation into specific communities and their social interpretation of these experiences. From a feminist perspective, people obtain gendered identity through participating in organizational practices, which guide their behavior to be manly and womanly (Gherardi, 1995). Gender identity is a product of sociocultural construction, which is influenced by people's experiences in different social contexts like early schooling, post-secondary education, workplace and home (Flannery, 2000).

As Wenger (1998) argues, the issue of identity becomes more individual because 1) each person becomes a unique intersection of forms of participation; 2) people have multimemberships of different communities, when people participate into different communities of practice at the same time and they get involved into multiple relationships through the experience of being one person; 3) people are situated in diverse social contexts where the identities are defined. Thus the process of learning is also a process of identity management, whereby people coordinate identities among multiple communities.

From a feminist perspective, people learn not only how to perform a task and develop their professional competences, but also they learn how to do it in a gender-appropriate way in communities of practice. However, if this community identity is opposite to their gender identity, they will confront dilemma situation in the process of managing different identities.

Gender identity and engineering identity

According to Lave and Wenger (1991), learning involves traveling along a trajectory from the periphery to the center and becoming a full member of the community. From a feminist perspective, this study argues that people's experiences are shaped by both the social group they belong to and the individual experiences. Therefore, men and women can have different access to participation into the community of engineering, in that the social image of engineer and the culture of engineering educational institutions are masculine. For the few women who enter a community where the prevailing norms are based on men's interest, they might have different experiences traveling from the position of newcomer on the way to center. Within communities of practice, there are attitudes, principles, and expectations providing messages in perspective of constructing a professional identity. There are also implicit expectations as to how femininity and masculinity should be done in local scale of community, which demands different experiences in the identity work and development process.

When using the term boundary, Wenger (1998) talks about how people can participate in different communities and obtain multimemberships at the same time. And by doing so, they bring different values and practices from one community to another. Besides, learning involves participation in a community of practice, but this participation refers not just to local events of engagement in certain activities with certain people, but to a more encompassing process of being active participants in the practices of social communities and constructing identities in relation to these communities' as well s an experience of meaningfulness (Wenger, 1998).

However, this study argue that whether and to what extent the boundary breaking can lead to active participation, full-membership and meaningful experiences is shaped and restricted by both the gender culture of the big scale of community and the prevailing gender relations in the local community of practice. In light of engineering education, there is only one valid identity of being an 'engineer' in this community of practice, which is labeled as masculine. Consequently, women are either excluded or have to sacrifice their femininity in order to adapt to the masculine culture when participating into the communities of practice, because the reproduction of values and behavioral patterns transforms them to be more masculine (Salminen-Karlsson, 1999).

In her research on gender in engineering education, Henwood (1998) confirms the general view of the marginalized situation of women in engineering education. She also witnessed the problematic identity construction process for women. She points out that seeing the identity formation of female engineering students as problematic implies expectation that there is a more 'normal' identity formation, instead of seeing that any identity formation is a result of the discourses in which it takes place. Rather than departing from an implicit problem and locating it, as has been done in previous research in gender in engineering education, Henwood observes how women are constructed and construct themselves in the different kinds of power relationships.

Kvande (1999), in her studies about female identities in engineering organizations, finds out four female identities to the organizational contexts which these women confronted with. These four identities result from different solutions to the problem of reconciling female values and engineering values, stressing the one and rejecting the other, rejecting both of them, or in some instances, being able to integrate them. A visible problem for these women was combining a feminine identity and an identity as an engineer at work. Whether they are able to manage this depends on factors like the organizational contexts, the environment of the community of practice at the workplace, and the strategies these women have employed to find their position and value in the community. The environment and values created for women differ from those for men.

Kvande's discussion about female identity construction in an engineering community of practice is also a process of handling the conflict of professional identity and gender identity. This is in consistent with Gherardi (1995)'s research on how women negotiate their identity and position being a 'traveler in a male world', which as observed, are quite different from how men do it.

Salminen-Karlsson (1997, 1999) and Stonyer (2002) also problematizes women engineering students' dilemma in face of the challenges of identity. Since the gender contracts as a whole define technology as a male task and engineering university as male sphere, women who have entered this area have been looked at as 'boundaries breakers'. These women are given a chance to get into the male world and show their capability and competence, but they are not thought of as real women any more, because as Stonyer (2002) argues, there are specific dominant socio-historical engineering identities within engineering which, given the traditions of engineering, are gendered as masculine. In this halfway position, the way to maintain the contract will be: to take it by adjusting to the norm at the price of changing identity, or to leave it to keep femininity in the normal social value (Salminen-Karlsson, 1999).

In summary, through the discussion of this framework, the concepts of gender and learning are related to the research context of engineering education. The general gender symbolism in western countries and the gendered (male-dominating) culture in engineering community establish a strong symbolic tie to link engineering identity with the hegemonic masculinity. Gender questions are embodied when women break the boundaries and enter this non-traditional profession (Du, 2006). The masculine culture in engineering communities of practice leads to differences between men and women concerning their experiences in the learning processes. This discussion also identifies how the engineering identity is developed in gendered forms through everyday practices, institutional activities and symbolic representations in micro settings. To conclude, gender relations play a role on both social structure and institutional culture, which directly influence the individuals' learning experiences. In relation to engineering education, what can be agreed from feminist literature in general is that this influence brings about more difficulties and barriers for women to study engineering than their male peers.

5.3 Possibilities for Change

One common concern in both the learning theories and the gender theories that have been discussed in this study is change. Learning both produces change and occurs as a result of it. Gender relations are undergoing changes through negotiation. Thus, this section will discuss possibilities for changes based on the framework of gender, learning and engineering education. However, this study argues that it is important to bring about the topic of power when taking about change. Because from the learning perspective, people are situated in communities of practice in which the dynamics of power and status are often controlling factors in how one knows and what one knows. From a feminist perspective, changes in gender relations are always struggles over the distribution of scarce resources in either material or symbolic aspects (Gherardi, 1995; Harding, 1996). Although men and women have different kinds and degree of power in different cultures, strategies for improving women's situations must always address issues of power imbalance (Hirdman, 1990). In relation to engineering education, modern science and technology have tremendous power to shape people's lives and identities, which makes gender a crucial issue in the development and application of science and technology (Berner, 1997). Therefore, in order to change the gender relations in engineering education, it is important to discuss the possibilities of changing the imbalanced power relations between men and women in science and technology.

5.3.1 Gender, Learning and Power

Learning and power

Power remains an outstanding issue when talking about knowledge, learning, and social relations. However, in this study I argue that power has not received considerable attention from any of the theoretical works on learning from the constructivist-sociocultural approach that have been reviewed in my study. Power remains untouched in the works of experiential learning (Kolb, 1984) and three planes analysis of participatory learning (Rogoff, 1995). To some degree, it is addressed in situated learning theory (Lave and Wenger, 1991) through the claim when looking at learning as a process where newcomers are travelling the trajectory from periphery to the centre, people are supposed to experience the process of gaining power. However, the concept of power is not really put into detailed discussion.

In his work on the theory of communities of practice, Wenger (1998) attempts to find a conceptualization of power that avoid both simply conflicting perspectives (for example, to perceive power as domination, oppression, or violence) and simply consensual models (for example, to perceive power as contractual alignment or as collective agreement conferring authority to) (1998:15). However, he provides neither clear elaboration nor proposal. In this way, communities of practice remain mutual, however, as argued by Tennant (1997), there may be situations where the community of practice is weak or exhibits power relationships that seriously hamper entry and participation.

Gender and power

From a feminist perspective, power is seen as a central aspect in gender relations in society, especially in studies on organizations. As both Hirdman (1990) and Harding (1996) point out, gender relations are always power relations.

Harding (1996) discusses different standards for scientific knowledge and ways different knowledge emerges through the politics of historical and cultural knowledge-seeking projects. She identifies how Western criteria for knowledge acquisition and validation tend to obscure women's contributions to knowledge. In the study of women's ways of knowing, the authors (Belenky et al., 1986) uncovered some salient themes — the experience of silencing and disempowerment, lack of voice, the importance of personal experience in knowing, connected strategies in knowing, and resistance to disimpassioned knowing. These suggest some hidden agendas of power in the way societies define and validate and ultimately genderize knowledge.

Perception of power in this study

When relating the discussion of power to the thoughts about Bildung, Masschelein and Ricken (2003) note that it is difficult to analyze power without the consideration of its historical transformation, specific context regarding time and place, and categorization regarding areas. Following Foucault's work, they pointed out that power is not only a certain type of relations between individuals, power relations are also deeply rooted in the social context. Power is not about direct action on others, but about actions upon their actions, upon the possibilities of their actions, both existing and coming.

According to Masschelein and Ricken (2003), in a traditional perspective, power is linked with terms of dominance, force, violence, repression, determination, etc. Opposite to it is freedom and obedience. From a postmodern standing point, it is necessary to think about power in terms of relations as a productive technique and mechanism. Power relations can be understood as productive processes of conducting different behaviors within certain possibilities as well as relations of individuals which operate through concrete forms of knowledge and discourse.

This study perceives power from two aspects. One is the dominance aspect of power. This study argues that power can be used to promote some groups and marginalize others; and not all individuals or groups can be put in the powerful situations and use their power for the benefit of the community. From a gender perspective, the different power distribution of men and women in society at a structural level will shape what can be learned by men and women and how they learn. If we look at engineering education in general as a community of practice, women's participation has been long hindered due to the male domination and men's mastery of power. Women were previously excluded from this community of practice, in earlier history by regulations and later by masculinity as the symbolic culture.

The other aspect is that power is not only suppressive or oppressive but also productive, as suggested by Masschelein and Ricken (2003). In relation to my study, power determines how individuals interpret their experience meaningful, and how the individual relates to him/her self to others through the social interaction in the learning process.

To take power into consideration provides different perspectives to understand both learning and gender in the individual-structural interaction. Firstly, from an individual perspective, it serves as a tool to see how individual identities are shaped by and shape the legitimacy of one's own experience when learning takes place. It also allows seeing how individuals manage their identity work through managing different identities. Secondly, from the structural perspective, it measures how one is recognized as competent, participant, being central/marginalized, it also measures whether or not and to what extent individual's ways of doing gender are appropriate in a specific organizational context. I argue that power is observed, produced and reproduced in the process of learning in communities of practice, which is also a process whereby individuals of both genders interact with others as well as the community, organization and social culture.

5.3.2 Changes

Change is an important factor at all the three levels of examining learning. Learning involves change when people become more experienced. Learning can be for the purpose of conformity or for change. People in contemporary societies live in patterned and organized relationships, and any change in one person will mean that the relationships are affected in some way or another, which will lead to the change of community. Thus, learning is itself one of the social processes that helps create the conditions for more learning.

From a feminist perspective, the classic social distinction of gender is undergoing change in that the social differentiation that gives rise to a sexual division of labor is no longer regarded as 'natural'. Therefore, as Gherardi argues, gender inequalities are becoming an embarrassment, and it is increasingly difficult in democratic societies to come up with moral justification for gender as a social destiny (1995:5).

To conclude, individual learning and doing gender, on one hand, furnishes the interactional scaffolding of social structure, on the other hand, individuals can, to different extent, through their learning process and doing gender, push social change both at the institutional and cultural level of social transformation and gender relation. Therefore, in this sense, both gender and learning as an individual practice and a social phenomena are created through interaction and at the same time structures interaction.

The emergency of the new is one of the central questions in the philosophy of Bildung as well. As Koller (2003) points out, knowledge is legitimated by being new or by being brought forth through the violation of existing rules. To conclude, these ongoing changes will unavoidably bring impact on the knowledge creation and dissemination process in engineering education. Individuals do not only take in knowledge and develop expected competences, but also participate in the process of creating new knowledge. The on-going changing process in both learning and gender relations will accordingly bring about changes in engineering education as well.

Relating the discussion of power and change in the integration of gender and learning to the research context, this study assumes that there is a mutual constructing process between individuals and the contexts they are situated in. This mutual shaping process involves transformation and changes of both individuals and community/structure. In this way, the learning activities in the engineering educational institutions do not only include the one-way engineering knowledge transition and technical skills training, which is based on the expected engineering competences. There is also the other way around that by studying engineering, learners are participating into the construction, reproduction and transformation of the professional identity of engineering community.

In light of gender, there are now societal changes which serve to promote women's access to the education, and the recruitment of more women into engineering education in the initiatives that have been taken did increase women's presence. However, taking the standpoint of gendered knowledge system as noted by Harding (1996), the growth of women's participation cannot guarantee an encouraging culture for women's learning in the community in that women might attain different knowledge from men if special values have been placed in the knowledge system and learning practice in favor of men.

Nevertheless, science and technology are not really neutral in the social and cultural construction and their relation with society is flexible, so, they are open to change. And according to Hirdman (1990), gender contracts system is a process of continuous defining and reinforcing expectations and behavior. The two sides in the gender contracts are continuously negotiating with each other, and the weak party can consequently get advantages step by step to be stronger (Salminen-Karlsson, 1999).

Summary

In this chapter, a model is proposed to link the concept of learning from constructivist-sociocultural approach and gender concepts from a poststructural perspective. Relating this integration of theories from two areas to the research context of engineering education provide a tool to examine learning and gender as both individual activities and social practices in the social context. This framework will function as an analytical tool for understanding both gender and gender relations in the research context to relate it to the empirical work.

The proposal of this conceptual framework as an analytical tool is a challenge to the theories of both areas as well as a challenge to the research context of engineering education, in which the focus of research has traditionally been focus on how learning functions as mechanism (from the perspective of learning) and how women's participation into non-traditional professions has been hindered and magilized. By proposing an analytical framework which focuses on the interaction of individual and structure as social phenomena, this study assumes a new mutual influencing processes between individuals and their social contexts from the perspectives of both learning and gender relations in engineering education. This is in tune with what Masschelein and Ricken (2003:151) intended to provoke: it is necessary to bring about discussions and thoughts about 'other categories and other ways of analyzing our present time in a pedagogical/educational perspective of permanent construction of sociality'.

6

Research on Gender Diversity in Engineering Education

This chapter reports the process of this research, from initiating the study, making decisions on paradigm location, choosing research topics, doing explorative study, constructing of theoretical framework, designing research methods, data generation and analysis process as well as writing. Also documented in this chapter is the researcher's reflection of the learning journey through carrying out this research: how strategies were developed when confronting the unexpected, changes, mismatch between literature on theories of methodology and the research context in my study, and so forth. Followed by is an analysis of the limitations and weaknesses of this research. Some epistemological questions that remained in the research are reported and related to a larger context for future work.

6.1 Main Characteristics of this Research

As Blaikie (1993) notes, research is a systematic enquiry with the aim of producing knowledge. In terms of the research aims, enquiries fall into three general categories: exploratory, descriptive and explanatory (Marshall and Rossman, 1999). An exploratory study aims to discover what is happening and searching for new insights and assessing phenomena. A descriptive study has the purpose of portraying an accurate profile of persons, events or situations. An explanatory study has a focus on explaining a situation or program.

Gender and Diversity in a Problem and Project Based Learning Environment, 99–133.

A research can have one of the aims or a combination of different aims. Seeking to bring about a better understanding of some social phenomenon, this study is close to both descriptive and explorative study. It is descriptive because the first aim of this study is to examine the practice of PBL concept from the students' perspective with respect to how they experience their learning process. This study is also explorative because it intends to find out what is happening to the learning processes of these engineering students of both genders under the influence of the learning environment as well as the prevailing gender relations in the Danish society.

6.2 Methodological Perspectives

This section will report the process of making decisions regarding where to locate this study with respect to paradigm as well as the choice making of qualitative study as the research methodology.

6.2.1 To Locate this Study in a Paradigm

A paradigm is defined by Guba and Lincoln (1994) as a basic belief system or world view that guides the investigator. In the western academic world, researchers feel it necessary to claim their adherence to one or another wider paradigm or worldview. In different literature the significance of identifying a paradigm and its relation to the instruments and methods in one's research is widely stressed as an imperative factor. The argument is that no research methods are self-validating so that the effectiveness depends on epistemological justifications (Blaikie, 1993). For example, Guba and Lincoln (1994) assert that questions of methods are secondary to questions of paradigm, not only in choices of method but in ontologically and epistemologically fundamental ways.

It is a difficult task to decide on a paradigm because of the complex and self-changing process of the terminology of paradigm, and because of the overlapping features among different paradigms. Before I move on to the discussion of whether I should and how to choose a paradigm for this study, and also in order to clarify the standing point of this research, I will provide a brief review of key features of the three main paradigms (positivist, interpretivism and critical theory or postmodernism) that have been observed in educational and sociological research by the end of the 20th century. These ideas are directly informed by the work of Guba and Lincoln (1994), and there is no intention of going in depth of the discussion about history, definitions and detailed comparisons of them.

(1) The positivism paradigm, which is characterized by terminologies of 'empirical', 'analytical', and 'quantitative'. Positivists believe that knowledge is discovered and knower and known can be separated from each other, thus they focus on experimental and quantitative methods to test and verify hypotheses (Lincoln and Guba, 1985). Positivists anticipate the time, context, and value free feature in generalization.
(2) Interpretivism, which is characterized by the terminologies of 'constructivist', 'hermeneutic', and 'qualitative'. Adherers of this paradigm expect multiple, constructed and holistic conceptions (Lincoln and Guba, 1985). In contrast to positivists, interpretivists seek to understand meanings through thinking and reasoning humans (Blaikie, 1993). This concern is extended by the constructivist argument that knowledge and truth are created and produced as the result of perspective (Schwandt, 1994). This paradigm also addresses the importance of context as well as the need of taking interests of different social groups into consideration in analyzing data. Given the concern with understanding members' meanings (Blaikie, 1993), interpretive researchers often prefer to use meaning oriented qualitative methods. They believe that the knower and the known are interactive and inseparable. They anticipate time, context, and value bound feature in generalization.
(3) Critical theory or postmodernism, which is charaterized by the terminologies of 'critical', 'postmodern', and 'neomarxist'. With skepticism, questioning and deconstructing as the basic doctrines, this paradigm is closely related to the concern with power, control, domination and inequality. Critical scholarship seeks to transcend and change the taken for granted beliefs, values and social structures that only benefit certain social groups. Postmodernists deconstruct and reveal the contradictions and exclusions of

> minority interests in the existing structures. In educational research, neomarxist evaluators seek to expose the 'hidden curriculum' and encourage to question and requestion the cultural, political and gender assumptions underlying the design of educational programs (Reeves, 1996).

As Lincoln and Guba (1985) states, each of these paradigms is a different, internally logical and consistent way of understanding the world. In spite of the contrast in epistemological beliefs, research representing different paradigms borrows research methods from other paradigms in actual practice. Each paradigm is advancing their premises and combining different insights in the self-sustaining process, and in many actual research studies use aspects of more than one paradigm (Gephart 1999). All these factors make the paradigm choice process a complex situation.

Categorization of paradigm and arguments among different paradigms make it rather difficult to establish a 'correct' academic credential in my study. The focus on learning makes it closer to the constructivism and interpretivism due to the concern with the subjective meanings — how individuals or members of a certain social group apprehend, understand and make sense of the social events and settings seeking understanding and interpreting human experiences. The 'what' and 'how' questions that have been asked in this study also lead to this approach which attempts to explore and construct the meanings of human actions.

However, current feminism study is often closely linked with post-modernism and critical theories due to their viewing society as essentially conflictual and oppressive (Lather, 1991; Marshall, 1997; Marshall and Rossman, 1999). Feminist philosophers challenge the traditional modes of knowledge production because they legitimize some but exclude other forms of knowing (Harding, 1987; Lather, 1991). Feminist researchers uncover cultural and institutional sources for oppression, and they name and value women's subjective experience (Marshall, 1997). With explicit emancipatory goals, feminist researchers often argue that research fundamentally involves issue of power and traditional research has silenced members of oppressed and marginalized groups (Rossman and Rallis, 1998). Therefore, gender is regarded as crucial for understanding both individual experiences and organizational practices.

This perspective provides a paradigmatic standing point to study gendered learning experiences in traditionally male dominated institutions in engineering education. It helps understand how female and male students perspective learning and interpret their learning experiences as meaningful through studying engineering. This perspective also helps answer 'why' questions afterwards by providing possibilities of taking cultural and institutional factors into consideration rather than only attributing gender differences in learning to biological and psychological reasons.

However, the attempt of combining theories of two areas brought about what is called complexity problem by Casti (1994) from the sophisticated philosophical thinking and the paradigmatic discussion. My research fell into confusion and puzzlement after the literature review in the areas of learning and gender studies. When I was struggling in the question of where to locate my research, Patton's (1990) work lit a light to my puzzlement. He questioned the well-accepted belief in the determinism of research paradigm on the methodology employed, and argues that the paradigm of inquiry adopted does not necessarily impose a fixed and rigid methodological choice.

Sharing similar ideas with Patton, Reeves (1996) developed a concept of 'eclectic-mixed methods-pragmatic paradigm' to argue for appropriateness in locating one's research in the research paradigms. From a perspective of research in instructional design, Reeves summarizes features of this paradigm in practice. It involves the openness of borrowing the methods of the above-discussed three paradigms in information collecting and problem solving. The mixed methods aspect relates to the recognition of the need for multiple perspectives to triangulate or bracket information and conclusions regarding complex phenomena. They view modes of inquiry as tools to better understanding and they don't value one tool over another. They also recognize the weaknesses of their tools, and are honest with themselves and their audiences about the tentative and probabilistic nature of the recommendations they make.

Inspired and supported by this mixed method paradigm, this study sees different tools are equally valid and valuable depending on how they are used. Modes of inquiry are viewed as tools to help with better understanding and they are only meaningful within the context in which it is used. The multi-methods paradigm provided a matrix for me to consider the most appropriate and feasible research methodology in this study.

6.2.2 Choice of Research Methodology[1]

Research is commonly categorized into quantitative and qualitative research in terms of what kind of data the researcher is seeking. Generally speaking, quantitative research tends to make predictions, use experimental methods to test hypotheses and to generalize findings and establish facts (Kvale, 1996; Hoepfl, 1997; Silverman, 2000). In contrast to the statistical nature of quantitative research, qualitative research is naturalistic, pragmatic, interpretive, multi-method in focus, and grounded in the live experiences of people (Denzin and Lincoln, 1994; Kvale, 1996; Marshall and Rossman, 1999). Qualitative research has become increasingly important and useful modes of inquiry for the social science fields, especially in educational settings (Hoepfl, 1997; Marshall and Rossman, 1999; Silverman, 2000).

I will not go into a detailed discussion of the comparison between these two research methodologies that have been argued on their relative values. Neither will I discuss more about the paradigmatic changes insides each of the research methodology as well as the recent advocation and encouragement of using both quantitative and qualitative methods to provide more convincing evidence (Hoepfl, 1997). What is intimately related to this study is the discussion of what Patton (1990) refers to as 'methodological appropriateness' in choosing and evaluating research methodology.

As stated by different writers (Blaikie, 1993; Marshall and Rossman, 1999; Silverman, 2000), it is dangerous to assume a fixed preference or pre-defined evaluation of what are 'good' or 'bad' methods for doing research, because there is no ideal way to gain knowledge of the social world. How to get a research done should depend upon many things: aims of the research, research questions, theoretical departure, understanding of the nature of the data, and the primary assumptions about the reality. Therefore, instead of trying to fit into a specific research tradition and following 'the' right approach, this study holds the belief that the choice of research methods should be research goal oriented in the designing process and contextualized in the empirical site.

[1] Following Blaikie (1993) and Silverman (2001), who distinguish the terminology of 'methodology' and 'method', I use the terminology of 'methodology' to refer to a general approach to studying research topics (how research should or does proceed), which include the choice about cases to study, methods of data gathering, forms of data analysis etc. In this study, I only employ broad definitions like quantitative/qualitative. The terminology of 'method' refers to specific research techniques.

A qualitative research methodology became the most practical and relevant option in my study for the following consideration:

Firstly, for the correspondence to the research questions and conceptual framework. The focus on understanding individual experiences in a social context as well as how learners interpret the meanings as well as appreciate the meaningfulness in the learning process makes it close to the characteristics of qualitative study. This study holds the belief that human behaviour is based on meanings which people attribute to and bring to situations, and that behaviour is continually constructed and reconstructed on the basis of people's interpretations of the situations they are in (Punch, 1998). Thus by focusing on experiences of learning, I intended to know the learning process of the engineering students of both genders through doing project work in groups in a PBL study environment as well as the factors influencing their learning.

Secondly, for the consideration of general small population in the research site, the engineering university in Denmark, especially the small participation of female student in one of the programs which were chosen for research (more explanation about this background and process will be provided in the later section). This makes it difficult to account for the generalization from quantitative method. Therefore, the consideration of using a combination of both quantitative and qualitative methods ended up with a negative decision.

Thirdly, for the feminist standing point. Qualitative research is especially a principle means by which feminists have sought to understand lives and situations of women in their everyday world (Lather, 1991; Kvale, 1996). Feminist research has at large a focus on women's diverse situations and the structures that influence those situations (Marshall and Rossman, 1999). Feminist researchers are concerned to explore and expose both the position of women and processes of the production and construction of gender relations in education at the level of the micro politics of educational institutions (Gordon et al., 2001). The aim is to make sense of women's experiences through interpretive conversations and actions in their lived world.

As Silverman (2000, 2001) asserts, there is no agreed doctrine underlying all qualitative social research. The variety of qualitative research[2] can be

[2]Definitions of qualitative methods vary. I found the description of qualitative research provided by Crossley and Vulliamy (1997:6) suits well with the design of my study: (Qualitative research) provides descriptions and accounts of the processes and social interactions in 'natural' settings, usually based upon a combination of observation and interviewing of participants in order to understand their perspectives. Culture, meanings

witnessed from different 'ism', as Denzin and Lincoln (1994) summarize the typology of qualitative inquiry, which are respectively constructivism and interpretativism, critical theory, feminism, ethnic studies, and cultural studies. Nevertheless, there are some shared preferences and prominent characteristics of qualitative research that have been identified by different authors. A list of these shared features as shown in the following is based on the works of some representative writers in qualitative research such as Lincoln and Guba (1985), Patton (1990), Denzin and Lincoln (1994), Rossman and Rallis (1998), Marshall and Rossman (1999), Silverman (2000, 2001). The purpose here is to gain a better understanding of this research methodology and clarify the links to my research.

Qualitative research are commonly characterized by

(1) inductive and hypothesis-generating rather than hypothesis testing;
(2) natural settings for data generation, which is from observation rather than experiments, unstructured or semi-structured interviews rather than structured interviews;
(3) qualitative and interpretative nature of data, which is based on the analysis and interpretation of words and images rather than numbers;
(4) interactive nature in the research process with focus on meanings seeking, in which the researcher uses self as instrument and reflect on his/her role systematically, and the viewpoint of the studied are taken into account and valued;
(5) valuing emergent process, flexibility, uniqueness and accepting changes in the research design rather than being predetermined and fixed;
(6) rich details of and insights into the experiences of the studied in the report.

These characteristics fit the aims of my research, which is to gain insights on social relationships and processes through discovery and description of individuals' practices, their response to their own actions, other's actions and the

and process are emphasized, rather than variables, outcomes and products. Instead of testing pre-conceived hypotheses, much qualitative research aims to generate theories and hypotheses from the data that emerge, in an attempt to avoid the imposition of a previous and possibly inappropriate frame of reference on the subjects of the research.

context. These features of qualitative research provide a wider range of possible materials for my study, because generally everything in the empirical work might probably become the source of data (Denscombe, 1998:221). Further, these preferences of qualitative research provide a direction and a framework for developing strategies and designs for the data generation, analysis and presentation in my research process.

6.3 The Overall Research Process

According to Marshall and Rossman (1999:25), in qualitative inquiry, the topic of interest can derive from different sources: (1) theoretical traditions and their attendant empirical research; (2) real world observation, emerging from the interplay of the researcher's direct experience, tacit theories, political commitments, interests in practice, and growing scholarly interests; (3) interaction of their personal, professional, and political interests. A combination of these three factors was witnessed in process of my research. A detailed description of the research process is provided as the following.

6.3.1 Explorative Study

In December, 2002, right after my master study, I started this Ph.D. project, which was under the umbrella of a gender equality-aimed project 'Get a Life, Engineer!' managed by the Danish society of Engineers (IDA). Based on my past experiences in both theoretical and empirical research[3] in engineering education in different social-cultural contexts, I had a background knowledge that when women worked in science and technology areas in many western countries, women, as boundary breakers, they were confronting identity challenges by feeling being 'others' (Harding, 1986) and 'guests' (Berner and Mellstrom, 1997). The main reasons were attributed to the masculine culture in the engineering institutions as well as the prevailing gender relation in engineering as a non-traditional profession for women.

Most of the research on gender and engineering education that has been carried out was focused on hard-core engineering disciplines and traditional lecture-based learning environment. It has been argued that women's learning

[3] A description of my background as a researcher as well as how this study is influenced by that background can be seen in the Appendix 8.

is better supported in an environment that is different from traditional ones, which are mostly male dominated ones (Hayes, 2000). In this light, PBL Aalborg model as a learning environment (as discussed in Chapter 2) can be, and has been supposed to be female friendly (Kolmos, 1991); nevertheless, with little empirical evidence. However, statistics did not show visible difference in terms of female participation from other engineering educational institutions in Denmark, where the learning environment is traditional lecture-based (women's participation in engineering programs in general is around 20–25% in both of the universities).

This background and contextual knowledge led to the initial question for my Ph.D. research: what are the experiences of female engineering students in the student-centred learning environment in a Danish context? In January and February, 2003, I conducted an explorative fieldwork at Department of EE Engineering at Aalborg University. Data resources in this explorative study can be seen in Appendix 3.

I started with an interview with the director of the study board, who provided me with both textual and oral information on the study environment of EE Engineering. When I mentioned the focus of gender in my study, he could not see the point of research on women's experiences based on the limited sample. And he attributed the invisibility of female students at EE Engineering to women's lack of interests in science and technology as well as the influence of primary education, 'the door of the university is open to everybody, it is just because women don't show interest in this area', as he said. Similar comments were heard in the later talks with other male lectures, supervisors and students.

Introduced by different colleagues at the university, I contacted 4 female students (Julie, Laura, Mia, and Clara) from Department of EE Engineering and one female staff member, assistant professor Patricia who works in another department but who used to be a student at EE. I explained to them the overall aim of my study and interest in their learning experiences. My special focus on women's experiences gained their interest and supports in my study. January is the exam month every year. With the permission of these students, their group members, supervisors and examiners, I observed two supervision meetings and eight oral exams.

Aims and focuses of these interviews and observation were 1) to gain a deep understanding of the PBL concept in the research context — how it is conducted in practice with respects to supervision, group work, and exams;

2) to have an idea of women's experiences of studying (hardcore) engineering in Denmark — how it is compared with what is described in different literature; 3) to ascertain research questions and design the following research in my Ph.D. project.

The observation part helped me gain a direct impression of the learning environment in the aspects of supervision and assessment. There was no noteworthy information relating to gender differences from the observation in this period. These female students performed actively in the oral exam, they gained above average scores compared with their male group members (Clara achieved the highest individual score among her group members), they got positive comments from their male supervisors and male group members (in the case of Clara because I only talked with the male peers in her group) in terms of their academic performance and achievement.

Interview questions with the female students and assistant professor Patricia were mainly focused on their past school experiences in terms of interest in technology, reasons leading to their choice of this engineering program, and their experiences of studying. In the story telling process, they brought about topics relating to gender by themselves, by referring to the hard time at the beginning in studying the hard-core subjects and at the same time getting used to working in an environment where the majority of people were male.

In the whole process of the explorative study, nearly none of the informants (except Patricia) thought of women's invisibility at EE as a problem. When asked about the question why there were so few female students at EE, most people answered that it was because few women had interest in science and technology. The terminology of gender equality was specially addressed by most of the male staff members I talked with. According to them, advanced level of gender equality has been achieved in Denmark, which allowed the room for women to do what they were interested in. Some male informants showed surprise to my question because for them gender equality was an out-of-date topic and professional choices were mostly attributed to individual choice based on interest.

This explorative research suggested that it is not an easy task to discuss the concept of gender only with the knowledge about the experiences of female students. Neither would it be convincing to analyze whether ways of learning is gendered without the knowledge about men's experiences. Therefore, the research focus was shifted to studying gender based on the experiences of both men and women rather than only focus on women.

Further, based on the recruitment experiences in other western countries, I started to rethink the general question of 'how to attract women to study engineering?' and 'how to educate women to be engineers?' Is the increase of women's participation into non-traditional professions equal to gender quality? Is recruitment enough? Based on these thoughts, the research interest turned to the topic of meanings and identity: 'what does it mean for women as well as men to learn engineering?', 'what does it mean for these students to be an engineering student and at the same time to be a woman/man?, and 'how do students of both genders appreciate the meaningfulness of the learning process in engineering programs in this student-centred learning environment?'.

The explorative study led to the consideration of possible research sites and informants. This process fostered analysis about what research questions would denote for practice. It also help proceed the construction of a theoretical framework.

6.3.2 Reframing Research Focus and Questions Through Constructing Theoretical Framework

The tendency of shifting the research focus became clear after the 3-month-long explorative study, that is, from examining issues on women's underrepresentation in engineering education to exploring the understanding of learning from a gender perspective in the research context of engineering education. Based on this shift of research focus, I established a model of understanding learning (in Chapter 3) and a model of understanding gender relations (in Chapter 4), which provides a comprehensive way of understanding learning and gender relations through examining the individual-structure relation in the social context. The establishment of a theoretical framework by combining these two models in relation to engineering education (in Chapter 5) provides an analytical tool for the analysis of empirical findings.

The theoretical frameworks proceeded the research process in the ways of (1) providing a tool to examine learning and gender as both individual activities and social practices in the context of organizations. (2) Helping locate the focus of this study on how learners seek meanings and develop identities through studying engineering as well as how they manage professional identity and gender identity. (3) Functioning as an analytical tool for the understanding of both gender and gender relations in the research context so as to relate to the empirical work. Tentative guiding hypotheses that were generated from the theoretical framework helped focus the general research questions, which, in turn, guided the choice and development of research strategies.

6.3.3 The Design and Development of the Research Practice

Once the research questions were identified, the choice of qualitative methods became logical because the questions required an exploration of an area of knowledge and practice. Research on little-known phenomena involves uncertainty and complexity, which demands openness to the unexpected findings and a flexible design of research practice with the consideration of contexts. The following sections will report the process of choosing and developing strategies in the research practice.

Employment of multiple research methods

Silverman (2000, 2001) summarizes four types of commonly-used research methods in qualitative inquiry: observation, analyzing texts and documents, interviews recording and transcribing naturally occurring interaction. As he emphasized that no research method stands on its own, in practice, these methods are combined for the purpose of triangulation. The design of methodology and choice of methods were based on the belief of appropriateness in order to link methods epistemologically to the focus of the study, the research questions and theoretical framework.

Multiple methods were employed in the data generation and analysis process. The focus on individual experiences suggests that the importance of understanding the meanings that people attribute to their actions, which involves deep understanding of their thoughts, feelings, beliefs, and values through face-to-face interaction. The expectation to look into the construction of the community culture from a gender perspective involves the presence of the researcher in the sites. Thus a combination of different methods like in-depth interviews, focus groups, and observation became the main resources of data. However, none of the employed method was solely existing and fixed, but rather applied flexibly based on the concrete research context. As Blaikie (2000) notes, the data collection techniques can be practises in a variety of ways, for example, interviews can be individual formed or group formed, and observation can range from total participation, to observation with various combinations, depending on the research aims, tasks and contexts. In the case of this study, the choice and development of different strategies were modified along the way due to the accessibility of informants, changes and new issues coming out in the research process, and the contextual differences among the study programs. The following texts will respectively describe the process of data generation and analysis in the two research sites of EE and A&D.

Choosing research sites

The Faculty of Engineering and Science, Aalborg University (AAU) provides 5 year (or 10 semester) — long programs which lead to master degrees. Students can become a member of the Engineering Society of Denmark (IDA) and are entitled to professional engineer by IDA since the beginning of the education. An overview of the structure and overall schedule in the faculty can be seen in Appendix 6-2. The PBL concept is implemented in all the study programs at AAU; however, they are practiced in various ways in the context of different study programs in different faculties. Two engineering departments[4] were chosen as research sites. First, Electrical, Electronics and Computer Engineering (EE), which can be seen as a representative of traditional hardcore engineering discipline. Women's participation as students and as teachers remains under 5%. Second, Architecture & Design (A&D), which is a comparatively new engineering discipline with the ideology of combining technology with creativity and design. The representation of women students and teaching staff are respectively around 60% and 30%.

The choice of these two research sites was based on the following reasons. 1) Based on the assumption that learning environment plays a role in the learning process of students, it is an intention to gain more knowledge about

[4]The organizational units at Aalborg University are rather unique. There are three faculties (Engineering and Science, Humanity and Social Science). Within each faculty, there are institutes (or departments as they are translated into English), which refers to research-based organization, and study boards, which refers to organizations responsible for arrangement and management of different study programs for education. Officially, at AAU, employees belong to institutes and students belong to study boards. In this study, there is no intention to get into discussion of this way of organization. In order to make the context of my study understandable for readers from both inside and outside of AAU, I use the terminology of 'department' to refer to EE and A&D as two engineering areas in this study. The main focus in my research is only on the distinguishing of EE and A&D as two engineering areas.

the students' experiences of studying engineering in a student-centered learning environment, compared with previous conducted research in traditional learning environment. 2) Based on the assumption that discipline matters in learning process, it was the intention to make a comparison between two different engineering programs. 3) Based on the assumption that gender plays a role in the learning process of students (especially with the evidence of feminist research in the traditional learning environment), it was the intention to examine the influence of gender relations on student's experiences in studying engineering in a student-centered learning environment.

Data generation

Access to the field at EE

In September, 2003, a second round of fieldwork started at EE. From the website, I searched the name lists of students in different specializations. I contacted the 9 female students out of around 400 students[5] by visiting them in their project room. Hearing the purpose of my research and the special interest in the learning experiences of female students, they (except for one[6]) showed immediate interest and agreed to talk with me about their experiences. I asked them to pick up the time and place that suited them. Among these eight female students, there were Julie, Sara, Mia and Clara, with whom I had developed a rather friendly relationship from the cooperation in the explorative study. With their agreement on time and places, I conducted interviews and follow-up interviews with them since late September to late November before the students entered the busiest moments for finalizing their semester project. Except for one female student who had the interview in her project room after their working hours, all the other interviews took place either in the cafeteria on campus or in my office.

With an agreement with Clare and her group members Christian and Carl, I decided to follow them observing their study life. This decision and agreement was based on (1) a closer relationship with Clara and warm support

[5] This number did not include the students in the first year and foreign students who joined the international program since the 7th semester.

[6] There was one female student who told me that she was in stress at the moment and preferred to be interviewed another time, though she never came back to me later and I did not push her by asking again. But she kindly introduced me to another female student who was her friend.

from all of them[7]; (2) my confidence in sensing the relaxed atmosphere in the group.

My original plan of interviewing the male group members of these female students was changed due to the ethical consideration. I asked these female students whether they would mind if I talk with their group members, they said that would be fine with them, and I could also interview their male peers either with or without their presence. However, I got a second thought due to the general situation of women as minority in the project groups. Eventually I decided to give up these chances of collecting potentially interesting data in order to avoid any risks that will possibly bring uncomfortable feelings, embarrassment, and potential barriers to the female students.

Instead, I chose some 7th semester (this is the time they choose specialization) male students at EE in order to compare with the interviews at A&D by choosing student at the same stages. This was also because most of the educational activities at this level are conducted in English. With the interest and permission from groups, I interviewed two (one with 4 male students, the other with 5 male students) groups and observed two supervision meetings and two group meetings.

Interviews

Data generation in both EE and A&D mainly replied on in-depth interviews as the primary method of data collection, which allowed me to understand perceptions of learning from the learners perspectives, as well as to what extent they interpret their learning experiences as meaningful. Qualitative in-depth interviews are much more like conversations than formal events with predetermined response categories (Marshall and Rossman, 1999). In practice, the interviews I conduced covered all the three types of interview strategies summarized by Patton (1990): informal conversational interview, the general interview guide approach, and the standardized open-ended interview. I exemplified a few general topics in hope to help uncover the participants' views with respect to paths to the engineering profession, past and current experiences in learning through doing projects in groups, perceptions towards learning resources, and opinions on the overall invisibility of women in engineering

[7]By then I had kept informal contact with Clara after two formal interviews with her. I met Christian and Carl when observing their activities in the explorative study and talked with them afterwards. And by then, they had continuously worked together in the same group for 4 semesters.

profession. In the process, I left room for the informants to frame and structure the responses.

The interview in my study was in the form of phenomenological interview, which is a specific type of in-depth interviewing grounded in the tradition of phenomenology. It suits the purpose of my research due to its focuses on the deep and live meanings that events have for individuals, assuming that these meanings guide actions and interactions (Marshall and Rossman, 1999). Following Patton's (1990) suggestion, in my research, three main phases were followed in the process of interviewing. Based on my own experiences as a researcher, 1) I developed and clarified preconceptions of the concepts of learning and gender. 2) Then I clustered the data around themes that describe the experiences and identified the essence of the phenomenon, 3) before I moved to the final stage by structuring all the possible meanings and divergent perspectives.

In the interviews with female students, I explained the background and purpose of my project, and how their information would be used and contribute to my research. I also asked their permission for recording, and all of them agreed without hesitation and they all seemed to be relaxed with the small machine. I used a mini-digital recorder to make it less visible so that it would reduce the chances that they might feel uncomfortable about the interview situation.

Interviews with male students were conducted in groups based on their homogeneousness and for the consideration of time and financial efficiency. The purpose of knowing about their group work made it convenient to gain information on how the group work was done.

Each interview began with some descriptive questions about their background, reasons for choosing this study program. Most of them were quite relaxed and talkative right from the beginning, except two female students who appeared slightly nervous for the first 10–20 minutes. Then interviews proceeded gradually at their own paces. The questions[8] flew from the reasons why they chose the engineering university, their past experiences of studying science and engineering subjects, the current project they were doing, to their spare time activities, hobbies, and relationships with family, teachers and friends, future plans, and opinions on some social phenomena, etc. Interview

[8] See interview guidelines in Appendix 6.

guidelines were designed as semi-structured form, but at the same time, I tried to keep the flexibility for the different interview situations. Interview situations differed from one to another, sometimes it went through according to the interview guidelines (for example in the case of a couple of female students who appeared to be shy); sometimes students brought about some topics in the interview guidelines without being asked (for example, all the female students brought about the topic of gender by themselves); sometimes some new topics and issues came out through the conversations (especially in the group interviews with male students, there arose different new topics when students reminded each other of things).

The interviews varied in length from one to three hours due to their differences in characteristics and their availability. Interview with the two male groups last especially longer because they sometimes went into discussion with each other when my questions reminded them of different past experiences. They also made jokes, told stories and showed me around their working surroundings and facilities. In most interview situations, I was mainly a listener when they got involved in their story-telling, meanwhile I would write down the main points and key phrases in my field notes. They were quite frank and open-minded to share their thoughts with me; trust and a pleasant relationship developed among the way. After each interview, I would recall the whole process and fill in details to expand the condensed account as soon as possible.

All the interviews were recorded and transcribed. In the case of individual interviews (the ones with female students), the preliminary analysis (some descriptive stories of their learning experiences) were sent to them afterwards for confirmation. Some of them responded immediately with confirmation, and some of them showed more interest in this activity and modified their stories (basically adding more thoughts and descriptions) before they returned it to me. I did follow-up interviews with more questions and discussions with those who did modification themselves.

Observation

From the research point of view, I was greatly interested in interviewing the male students and observing the daily practice of doing project work in groups after the interview with the female students. Because observation is a fundamental method in qualitative inquiry that entails the systematic noting and recording of events, behaviours and artefacts in the social setting chosen for study, and it is often used to discover complex interactions in natural social settings (Marshall and Rossman, 1999; Silverman, 2000; Flick, 2002). Participate observation was originally designed to be the main resources of data due to the opportunity this methods offers to learn directly from the first-hand involvement in the social world based on the researcher's own experiences (Spradley, 1979).

However, there was a potential risk that my presence might interfere with the naturalness of the events by bringing about tension to the female students, because most of them were single female student who worked with 5–6 male students. This concern went particularly for those who talked about their 'problems' with me and who were still struggling to 'fit into' the culture. With this concern, I eventually decide not to take the risk and to adjust the research design. This decision was also based on the rich data I had gathered through interviews. As a result, observation at EE was mainly uses as strategies to understand the learning practice of studying engineering from the students' perspective.

Data gathered from the observation[9] with Clara's group turned to be a holistic description of events and behaviour. The early stage observation was mostly based on broad interest in their daily life without predetermined categories or a strict observational checklist. Later on, I identified certain patterns of practice in terms of learning process, learning styles and social activities (communication and interaction, for example). From late September to December, 2004, I followed their group to observe their lectures, meetings with supervisors, lab work and group work in the project room. The frequency of my attendance ranged from 2–3 days per week to participation into special events. My focus was mainly on the main events in their daily practice, the process of their project work in terms of planning, organization and management. I made field notes and asked them questions along the way. The main purposes for the observation were 1) to examine the practice of PBL from an angel of students' experiences; 2) to enrich as well as triangulate the data generated from interviews.

[9]Marshall and Rossman (1999: 107) categorize observation into different types: highly structured; detailed notation of behaviour guided by checklists; and more holistic description of events and behaviour.

Access to the field at A&D

September, 2003, I tried to use similar strategies (as I did at EE) to access students at 7th semester[10] at A&D, but without success. I could not find the student name lists on the website, and I had to turn to the study secretaries for 7th semester. The responsible secretary asked me to write a description of my project, which she could send to the all the students by email and those whoever would show interest could contact me. I accepted this way though I preferred to reach contacts face to face. Unsurprisingly, this way of access did not work well — for more than one month I did not get any reply in this second-hand method. Fieldwork at A&D was postponed. I had to modify my research plan, due to the time limitation of the whole project.[11] I decided to adjust the time distribution of fieldwork from conducting two two-semester-long fieldwork at two research sites at the same time to conducting two one-semester-long project in autumn semester in 2003 and spring semester 2004 respectively.

A new strategy was developed for the second try in accessing A&D. At the end of year 2003, I contacted the head of the department, who heard about my research before and showed interest in it. With her introduction, I accessed the head of study board. I had an interview with him, with a focus on the history and development of this department and study program. With his great interest in my project, I got warm support to start my fieldwork in the 6th semester students from the coming spring semester. This choice of students was made because they just started with their specialisation study at the moment. I also accessed different responsible people for the students — the coordinator and the secretaries. I got special entrance key, which allowed me to enter all of their buildings for different activities.

At the beginning of February, 2004, I made a presentation of my project in the first formal gathering for the 6th-semester students in the specialization of industrial design. My project raised the interest of students due to the concern with their learning experiences. With their agreement and support, I started my fieldwork with observation with their group formation, mini project exam, topic choice, lectures, and I followed two groups, which use English as the

[10] I intended to choose students at 7th semester as the main target because it was the moment they had chosen their specialization and they might have quite something to tell after over 3 years' experiences.

[11] My Ph.D. project was financially sponsored by the project 'Get a life, Engineer!' from December, 2002 to May, 2005.

language for supervision meetings (because their common supervisor was a foreigner) and some group discussion.

Explorative observation

The fieldwork at A&D last a semester (from February to June in 2005). The early stage observation was mostly explorative with broad interests in their daily life and I tried to figure out the options for focus. In the beginning two months, I followed most of the activities for all the students (whatever accessible and informative) in order to gain more knowledge about this discipline and study program. When observing lectures, I was mainly sitting there observing and making notes. In some lectures, I was sometimes invited by the lecture to participate into their activities, for example, drawing. When observing supervision meetings and group meetings, the focus was on the proceeding of project work, the planning, organization, and management process, and the interactional patterns.

Group interviews

Based on suggestions of different literature (Krueger, 1994; Litosseliti, 2003), group interview was used as the primary resource for data generation, and have been found useful in the following aspects. 1) to reveal the beliefs, attitudes, experiences and feelings of the informants; 2) to obtain a number o different perspectives on the same topic; 3) to examine participates' shared understandings of everyday life, and the everyday use of language and culture of particular groups; 4) to observe group dynamics and gain insights into the way in which individuals are influenced by others in a group situation.

However, limitations of this method have been warned by these writers as well (Krueger, 1994; Litosseliti, 2003). For example, 1) the possible bias and manipulation; 2) False consensus. 3) Difficulty in distinguishing between an individual view and a group view. 4) Difficulty in analysis and interpretation of results and making generalization.

Nevertheless, I found this method the most suitable option that can be applied to the research context at A&D. In addition to the above mentioned strengths that fit into the aims of this study, it was also based on ethical considerations. The evenly distributed gender participation reduced the potential possibility of making minority part uncomfortable in a situation of group discussion. This is different from EE, in which interviews with the female students were conducted in the individual form due to the consideration of their being minority in the group.

Keeping both the strengths and weaknesses in mind, group interviews were planed along the way of the explorative observation. In the middle of the semester, I interviewed 5 of the 8 project groups and one female single-sex group in the specialisation of architecture design. Among all of the 6 focus groups (with about 30 students altogether) I interviewed, there were 3 female single-sex groups, 1 male single-sex group, 2 mixed groups. The group size ranged from 4–6 people in each interview. All of these interviews took place in their project room in the open studio. The interview questions (see Appendix 6) were planned based on the explorative observation and the questions that had been asked to the students at EE.

Focused/selective observation

Due to the easy access (all the groups were situated in the same open studio), I followed different groups with their group work and supervision meetings. Later on, I identified certain patterns of practice in terms of learning process, learning styles and social activities (communication and interaction, for example). My focus was reduced to one group after one month. Group 1 was chosen by chance (from an informal talk) due to their interest, their choice of English as working language, and having both female and male students in the group. Observation of the daily routine activities generated detailed information about the learning experiences of students. Combined with data from interviews, informal talks, questions along the observation, this helped to gain understanding of their opinions and attitudes towards different activities and events.

Data analysis

The multiple methods employed in the data generation process as well as the continuous modification due to changes in the context resulted in a messy, ambiguous, and time consuming process of organizing it and interpreting it with neat themes. As Marshall and Rossman (1999) note that there is no fixed and standardized strategy for data analysis in qualitative inquiry, the whole research process was also a process in which I developed analytical strategies along the way.

Following Kvale's suggestions (1996), data analysis process starts from the beginning of the data generation process. In my study, data analysis went hand in hand with data generation in order to interpret data in a coherent way. A guideline of data analysis was developed based on the preliminary research questions and theoretical framework, which was reflected in the tentative guiding hypothesis and was used to analyze the first interview data.

The process of data generation was also the process of analysis and data validation. After the interview of the first interviews, I listened and re-listened to the records and reflect on the research questions as well as theories. These reflections and questions to myself as well as to the former informants were written down in the research diary.

In order to make more sense of the data, I transcribed the complete interviews (including both individual and group interviews) on my own rather than using software programs. This choice was for the following concerns: 1) consideration of confidentiality; 2) consideration of validity — to reduce the chances for misunderstanding due to language issue — as the interviewer, I knew the context knows the contents of discussions; 3) to gain more intimate understanding of the contents of talks, to reflect upon the interview situations, and to carry out analysis along the way by relating to research questions and theories. The process of listening, re-listening and reading, re-reading led to familiarity with data and deep involvement in the research, which was of great help for me to edit them. This is also a process in which I develop analytical strategies by generalizing categories, themes and identifying patterns.

The interviews were transcribed word-to-word in the first version to be kept as a record. All the information that might potentially involve privacy or indicate the identification of individual informants (especially in the case of female informants at EE) was deleted. Following the suggestion of Kvale (1996), in the places where it is needed, modifications were made from oral language to written language so as to provide fluent written text for reading.

New questions and viewpoints came out along the way, which helped to move on and develop this research. This study found that discussing with informants about the preliminary analysis and asking confirmation as well as new questions in the follow-up interviews was an efficient way to validate and increase the reliability of the data. This is also a mutual-construction of the knowledge by both the researcher and the informants. For the individual interviews, transcripts of the tape record were sent to the informants for confirmation. All of them made correction and modification, which were mostly language check (from oral language to written language). In this process, there was correction regarding language as well as different new stories adding in, but no case of changing opinions.

Technically, I could have used computer-based programs for data analysis to make it more systematic. However, the empathy and devotion as a researcher made me get more and more involved and get through different round of analysis process on my own. Through relating the data to the literature review and theoretical framework, I gained deeper and deeper understanding of my research, the thoughts of which were kept in the research diary (with the reflection thoughts and questions in the research process). Group interviews were mostly checked up and validated through follow-up questions to groups or individuals.

I used color coding of notes to keep the track of dates, names, titles, attendance at events, chronologies, descriptions of settings and so on. This was useful to piece together patterns, define categories for data analysis and write the report. However, the interview transcripts, field notes, and research diary turned out to be nearly one thousands pages, which involved overwhelming work and many chaotic moments in the process. The process of reading, interpreting, member check, and reflecting on the theoretical framework led to more questions to the data and developed new understandings of data, which in turn, brought about needs for new data generating and analyzing. This process involved different questions, new ideas and new understandings coming out, which demands alternative explanations and changes in the research design along the way.

In this way, the data analysis process did not proceed linearly as expected, but rather, it involved many backs and forths. However, it turned out to be exciting and intriguing, especially when I sometimes looked back to it, I might get some new thoughts. Thanks to the patience and support from most of the informants, who provided easier access for me with patience and informative replies in my follow-up interviews.

Writing the report

In the report, pseudonyms were used in transcripts and report writing to protect the anonymity of informants. The informants were promised every reasonable attempt to maintain confidentiality.[12] The discussion of the findings fell into three parts. First, data generated during empirical work through interviews and observation was presented respectively in chapter 7 and 8. In Chapter 9, the description of the findings are summarized and compared between the two cases, before they are discussed in relation to the theoretical framework. This discussion leads a way to the answers to the research questions.

There are tensions and inconsistencies in the research process when I learned more and more about the engineering culture, which turned out to be visibly different from the social science culture where I came from. Difficulties were confronted in the research in making the gender issue visible to male engineers. Therefore, I was put in a dilemma situation facing different potential 'audiences' of my study. As Steier (1991) notes, the conversation with the research subjects and the conversation with colleagues are different conversations and different language games. Here, the researcher plays a role of translator between the two cultures, though it is never possible to make an exact translation. With the awareness that engineers and academic people from social science hold different beliefs in science and knowledge creation, as well as different expectations of the way I present my research results, I made an effort to try to make sense of this study for audiences from both cultures. Based on the suggestions by Eisner (1991), I structured this report by using themes derived from the data as focal points, and provide 'thick' description relating to the themes before summarize them at the end. I hope that by doing this, it is a way to meet the need for adequate description of the phenomenon from social science culture and the need for exact and logical structure from the engineering culture.

This report turned out to be orthodox based on my belief that some certain conventions should be fulfilled. This intention is to make it understandable and acceptable for the potential audience who might be rooted in different research paradigms.

[12]When accessing data, I did not bring any consent forms to the students due to the stronger belief in the mutual trust than signature. This issue was discussed with most of my informants with respect to their feelings about the different ways of handling confidential issues. None of the students in my research claimed that they favoured a contract.

6.4 Methodological Reflections

The research progress witnessed complex issues like choice of multiple methods, changes of design based on unexpectations in the context, lots of use of the self, influence of my biography,[13] empathy and sensitivity as a research, and so forth. A question is raised towards the context-related research methods: in which way it is qualified and scientific research? A reflexive discussion is followed based on the criteria about the quality of a qualitative research suggested in different literature (Eisner, 1991; Silverman, 2000).

6.4.1 Role of the Researcher

To borrow the 'traveler' metaphor[14] which Kvale (1996) uses to refer to interview researcher, I see the process of my research as a journey where I as a traveler explored knowledge in some certain area by visiting other people's lives.

Being a traveler in an explorative journey, I showed interest and respect, to the people I visited. The efforts of developing good relations through interaction and building trust helped ensure access to the interested events and people as well as their perspectives and insights of interpreting these events. By being present in the sites, I entered into the lives of the people I visited. Through asking questions, conducting conversations, and observing their life practices, I gained knowledge about their ways of living and their opinions, attitudes, beliefs in their lived experiences. By traveling through different settings and visiting different people, I gained new knowledge and insights, which provided different perspectives to examine the knowledge I had in mind before and to create new knowledge based on the past knowledge and past experiences as a researcher.

Both as a traveler and a visitor, the self of the researcher plays a role of instrument in this journey, which is a fundamental to qualitative research (Kvale, 1996; Marshall and Rossman, 1999). The deployment of the self drives

[13] A biography is attached in Appendix 8.

[14] A 'miner' or a 'traveler' are the two metaphors Kvale (1996) uses to illustrate the implications of different theoretical understandings of qualitative research. The miner metaphor suggests that knowledge is waiting in the subject's interior to be uncovered by the miner. And the traveling metaphor refers to a postmodern constructive understanding of the creation of knowledge.

the research to move on and also brings about a range of strategic, ethical and personal issues.

Throughout this chapter, I have discussed how deploying the self in my research helped develop strategies to gain access to data, generate rich data, and reflect on the theoretical framework with different new insights and perspectives as well as achieve deep understanding of the research. This turned out to be a mutual influencing process.

Also has been discussed in this chapter was how ethical consideration influenced the strategy development and adjusted methods choice in different contexts in the research process. Qualitative research deals with sensitive, intimate and innermost matters in peoples' lives, and ethical issues inevitably accompany the collection of such information (Spradley, 1979, 1980; Punch, 1998; Kvale, 1996). Following the three guidelines discussed by Kvale (1996), that is, informed consent, confidentiality and consequences, ethical consideration was kept in mind throughout the whole process of research from the matizing, designing, fieldwork, interview transcription, data analysis, to reporting and publication.

The aims of the research were explained to the students, most of whom provided warm support and inspiring suggestions regarding access to data. Some informal interaction with the informants helped to create a more relaxed atmosphere which led to more informative and rich data. In the interview and observation process, the importance of the researcher as a person is magnified because of the self used as the main instrument. On the other hand, this research experienced the difficulty in finding appropriate ways of using the self.

This research also exemplified the importance and difficulties in balancing the paradox between avoiding being intrusive to maximizing the 'natural' feature of the setting and at the same time building trusting relations in association with generating rich data. In some settings and events, for example, experiment work in the labs, group meetings and supervision meetings, although I intended to be the neutral observer, who engages not at all in social interaction, I might have touched the 'natural' settings by merely being there. However, in none of the situations students seemed to be bothered by my presence, instead, with the consideration of my limitation in the Danish language as well as technical knowledge, they explained to me what had taken place. When asked whether they got bothered with it, the answers were normally no, because they would not do it (explaining things to me) if they did not have the confidence in managing it. Before the exam time, my observation became more 'silent', as students turned busier with their project work.

6.4.2 Reciprocity and Power of the Research

In qualitative research, people who participate in the research are more than subjects to be researched; rather, they are involved in the process of constructing meanings and knowledge together with the researcher, which is also a process with power issues (Mishler, 1986; Kvale, 1996; Marshall and Rossman, 1999; Silverman, 2001).

This study has been empowered by the support from all the informants by the openness and willingness to share their experiences with me, their interest in my project, and their hope to see the outcome of this research. Some of them provided suggestions on possible events that might be useful for my research, introduced me to other possible informants, and offered opportunities for me to enter into their daily life. What have been achieved from them were not only rich data, but also inspiration and motivation to precede this study.

One question that I kept in mind along my research process was: what can I do for the participants of my research in return?

The most important thing I tried to do is to finalize this report as a gift to those who, by any means, contributed to this project. By providing a picture with the lived experiences, which was interpreted from the perspectives of themselves and me as researcher, I hope that my work can bring about reflection, rethinking and potential positive changes in promoting a more friendly learning environment. As Phil, the coordinator at A&D said, 'I am looking forward to see whatever coming out of your research, which will be another perspective to reflect and evaluate our work here.'

Therefore, this study is in hope to empower students of both genders by analyzing their active learning process in which meaningfulness of learning can be (has been for many students) appreciated, and by valuing their opinions, experiences and knowledge. As Litosseliti notes, 'the emphasis on participants' own language and the process of developing and re-framing their views as a result of collaboration can be important to the participants themselves' (2003:19).

This research is also an attempt, through the analysis of the learning process from a gender perspective, to bring more confidence and motivation to females engineering students. As one group of female students at A&D said, 'we are working hard to get through this high level education, and we want to prove that we can be as intelligent as guys, we hope that your research will prove this with strong evidence.' Further, my assistance with arranging and organizing the intercultural workshop between students at A&D and Chinese students in Beijing was another effort for me to enrich and empower each other.

After interviewing female students at EE, I encouraged them to modify interview transcripts to make them fully accurate; and I regarded it as a way to share power with the informants. This process, according to them, turned out to be a way to reflect upon themselves and get to know more about themselves.

The ethical consideration around intervention sometimes created personal dilemma during my fieldwork. During the courses of interviews, a few of the female students at EE talked about their lack of confidence in some hardcore subjects and the hard time in struggling to catch up their male peer group members. Each time at this moment, I became so empathically involved in the situation that my appreciation for their trust as well as my personal impulse encouraged me to give them some supports as friends. However, the rationality of being a researcher reminded me that I should continue the interview to keep the flow going. So very often, at the end of the interview[15] (after I turned down the recorder) we had friendly chats and discussions, sharing personal experiences of how to handle the situation of being a newcomer in unfamiliar situation and how the learning process in this situation led to change and person growth. In all the cases, these discussions led to warm and joyful mutual communication. By doing this, I also hope that this research would empower them by providing an opportunity to voice their experiences, feelings and thoughts in a safe setting.

During the courses of interview with male informants at EE, sometimes my impulse from a feminist perspective raised when hearing the biased words towards female engineers. However, I never allowed myself to comment on their unconscious opinions and to argue with them from the perspective of either a feminist researcher or a female person. Instead, I responded with patient ears as well as open questions. There was no argument with any of the male informants in terms of the topics on gender. In the situation

[15] I did this with most of the interviewed female students, except two of them with whom I did not manage to promote closer and experience-sharing conversation after the interview. I attributed this to personality reasons in developing communication..

of interviewing male staff members, my humbleness helped to gather more insightful information. In the cases of interviewing male students, my probe questions and their general awareness of reflection made them rethink about the reasons of some events in their past experiences. For example, when two male students talked about the dropping-out stories[16] of two female students in their group, they firstly used the joking tone that they 'managed to frighten two girls away in two months', followed was a serous explanation that 'the study was too harsh (high level) for them'. After my questions 'how do you think about it right now when looking back?' they said that when they rethought about it, it was partly because of the subjects, partly because of the male students, who took things for granted and made it difficult for the female students to participate.

I appreciate the slight changes that my research made to these male engineers, which might potentially bring about bigger changes in the future. I also appreciate the positive attitudes of some male students who see the value of their female peers for both their professional intelligence and social capability. However, as a young researcher, I still felt powerless confronting those male engineers who had deep-rooted beliefs in terms of the gendered labor division in society. I did not put much weight in my writing about this,[17] partly because the consideration of protecting the feelings of the potential female engineers readers of my report, partly because of my naive intention of keeping hope for further positive changes.

6.4.3 Limitations of the Study

Patton (1990:162) notes that there are no perfect research designs and there are always trade-offs. This study is no exception. Limitations of this study are reflected in the following aspects.

(1) This study put a special focus on people's understanding of meanings in their lived world by describing their experiences and self-understanding. Even though multiple methods have been employed for data validation and reliability. This study can still be charged of being especially subjective. Some other strategies

[16]This story is written in Chapter 7 with quotation.

[17]Here I refer to those who had negative opinions on female engineers with biased words.

could have been used to make this study more convincing to readers from some different paradigms, for example, combining both qualitative and quantitative methodology. However, by focusing on uncovering and describing the participants' perspectives on events, this subjective view is what matters.

(2) The use of the self as an instrument in a qualitative research can be a benefit and the same time a drawback. It will help obtain new findings in some situations where the informant's behaviour is influenced without being recognition. This will be beneficial to enrich the data collection but it might create potential weaknesses of poor credibility and reliability. The interaction between the researcher and informants can also be problematic, in that the story telling process and the replies of the informants might have been shaped based on what they thought I wanted to hear, or based on what they thought that they might be.

A good relationship can lead to appreciating cooperation and rich data, while overemphasis will bring problems of ethics and bias. In the research process, a good relation was developed with most of the informants, which directly helped to the generation of rich data. Students told me their life stories, which provided me rich data and text for interpretation. Through writing this process in the report, I intend to express my voice and their voice. Nevertheless, both the question and interpretation were still mostly from my perspective, that is, a female and feminist researcher. The question remained was: whether and to what extent the gender of the research matter? What kind of knowledge will turn out of this study if the researcher is a male researcher from another background than me?

By seeking to understand and interpret how individuals understand and interpret their learning experiences, this study also raises questions of what is called intersubjectivity (Gephart, 1999), which refers to the interplay of subjective, objective and intersubjective knowledge. This intention and process of knowing other's mind has been under debate in terms of how we know others' mind (Gephart, 1999). A question or challenge comes out of this study is whether what we have known is what has been known.

(3) The different methods chosen for the two research sites brought about difficulty in making comparisons. However, this was based

on the consideration for adequate information, cost-effectiveness, feasibility in terms of the subtleties of the setting and resources available for the study at that moment. It could have been done in another way if I chose some cases or another time for data collection. Complementary mixes of different possible approaches were used in the research process. In additions, choice of reduction was made when facing huge amount of data coming from multiple methods, however, there might have been some 'wrong' choices that some otherwise useful data were ignored.

(4) There are possible drawbacks in each of the concrete research methods.

— The semi-structure facture of the interviews benefits increasing the comparability as well as the structuration of the data. Nevertheless, there is the potential danger of obscuring rather than clarifying the subject's viewpoints through the different types of questions. This requires the researcher to have a high degree of sensitivity to the concrete course of the interview as well as the interviewee so as to make immediate decision (Flick 2002). I hope that this danger might be reduced by having a great deal of overview of what has already been answered and how it is relevant to the research questions.

— The phenomenological features of the interviews have the advantage of permitting an explicit focus on the researcher's personal experience combined with those of the informants. However, it is labour-intensive and demands a continual reflective process for the researcher.

— In this study, the method of observation mainly functioned as a way to make explorative study before the interview and to check up for the validation of data generated from interviews. With these as focus, it helped to increase the quality of the interview data (Kvale 1996, Litosseliti 2003), however, there might be chances for missing interesting new ideas and ignoring informative data that will lead to some other new knowledge.

(5) Qualitative approach in general is often criticised that it coexists uneasily with generalization, representation and transferability. On one hand, as has been claimed, this research is bounded and situated in a specific context, which means that the usefulness for other settings should be reassessed and judged by the readers. On the other hand, there is still an intension to take the risk to indicate the broad applicability by illustrating particular social phenomena. I could have interviewed more male students in the program of EE in order to raise the reliability with more evidence for my conclusion of the construction of masculinity in the hardcore engineering culture. In the program of A&D, I could have conducted more individual interviews to raise the validity. However, as Patton (1990) states, there are no strict criteria for sample size in qualitative research. This is agreed by other writers, for example, Eisner (1991) has similar notion that there is no statistical test of significance to determine whether the results count because of the employment of multiple forms of evidence in qualitative research. What information will be most useful and what information will have the most credibility are judged by the researcher and the reader (Hoepfl, 1997).

However, a support can be found from Kvale (1996), when he argues that the way it works is to obtain significant knowledge or consistent patterns for larger groups through intensive case studies with a few informants. He suggests two grounds for this: quantitatively, there are massive observations of single individuals in each case; qualitatively, the focus on single cases makes it possible to investigate in detail the relationship between a specific behaviour and its context, to work out the logic of the relationship between the individual and the situation. In addition, Mellstrom (1995) provides some beliefs on imagination. First, small-scale events in everyday life can illustrate features of broader social processes; second, fragments of recorded talk, extracts from field notes and reports of observed actions can reliably represent a social world; third, people make sense of their everyday lives, thus the complexity of the world can be understood through analyzing their descriptions and accounts.

(6) The use of English language as research language can bring questions towards reliability and validity of data. In the research process, communication between the researcher and informant went on

well in English, due to the fact that English langue is popularly used as a second language in Danish society, and in particular, English is well-used as the operational language at the master level program of the university. Nevertheless, different strategies were employed to confirm that information was correctly transferred from the informants to the researcher. This was mainly conducted through different ways of check-up: (a) in the interview process, asking probing questions, asking confirmation of the immediate interpretation of informants' statement. An example can be seen in appendix; (b) discussing preliminary data analysis and asking following up questions after the interview; (c) observing and informal talk. Due to the limitation of my Danish language, there was some otherwise valuable information that could contribute to the research. For example, how students argue in the group work, gender language, to check up what they say in the interview and what they say in the daily practice.

(7) My role of being an 'outsider' of both Danish culture and engineering culture can be both strength and weakness. There are many things I can see with the foreigner's eyes in the Danish culture and engineering culture, which might otherwise be taken for granted by 'natives'. However, there might also be some information that I am not able to catch due to the lack of knowledge in some specific areas.

Summary

In this chapter, I have reported the whole process of this research: locating my research in a paradigm, initiating research topics, explorative study, establishing theoretical framework, designing research methods, data generation and analysis process as well as writing. All in all, looking back the research journey, many pitfalls can be seen from self-reflection. Many things could have been done in another way, which might generate some different knowledge than it has been shown here. The research is also a process whereby the researcher applies what literatures suggest to the real field. Relating the literature to the context in my research is also a process whereby I as a researcher seek meanings and learn along the way. The whole research process was also a journey for the traveler, visitor, and research to learn and develop. No one approach or strategy provides a perfect solution for the researcher, because all methods involve assumptions, judgments and compromises, and all are claimed to have deficiencies. However, as Blaikie (1993) notes that, depending on where one stands, it is possible to argue their relative merits. Through the discussion of the whole process of my research, I hope I have given my readers the tool for evaluating the following text in relation to the construction of my study.

7

Studying Electrical and Electronics Engineering

This chapter reports the empirical work in the department of Electrical, Electronics and Computer Engineering (EE). The description starts with a brief introduction to the department, and continues with students' choice of engineering and the specific program, their experiences of being newcomers. This is followed by is the report of their perceptions of learning, and the learning processes of studying engineering in the PBL environment through doing projects in groups, attending lectures, gaining help from supervision and participating in different social activities. This learning process also witnesses a development of engineering identity through the curriculum facilitation, developing engineering ways of thinking, and interactions and communications. Also described in this chapter is the culture of the study environment in a gender perspective regarding gendered proportion, gendered features in the learning process, and the learning culture of this engineering program.

7.1 A Brief Introduction to the Study Programs in EE

This section provides brief introduction to the Department of Electrical, Electronics and Computer Engineering with focus on its origin, learning objectives, and curriculum design. The data resources come from the following sources: relevant documents and website[1]; interviews with the head of the

[1] See http://www.esn.aau.dk/

Gender and Diversity in a Problem and Project Based Learning Environment, 135–200.

study board, who had a background of electrical engineering himself; interview with five male teaching staff who worked as supervisors of the students in the explorative study (see Chapter 6 and Appendix 3); interview with four female teaching staff who worked in this department after they graduated from this study program.

The Department of Electrical, Electronics and Computer Engineering (EE) has its origin from two technical colleges[2] which were merged into Aalborg University when it was established in 1974. As one of the eldest institutes, it provides 5-year-long study programs which offer a Master degree in science and engineering (M.Sc.Eng). The programs contains 7-semester-long Bachelor level study programs, which are held in Danish and 4-semester-long Master level study programs, which are held in English and open to international students.

Curricula in the study programs in EE (see Appendix 2) include two semesters (the first academic year) of basic core courses; three semesters of advanced core courses and five semesters of specializations. The basic courses are provided in the 'freshmen building,'[3] which covers elementary skills and science knowledge in maths, physics and electronics as well as study methods. Students at 3rd–6th semester are expected to acquire a wider knowledge and professional competences in electrical, electronics and computer technology. Since the 6th semester, students will go into advanced study in specialized engineering areas such as control engineering, communications network, telecommunication, electrical energy technology, signal processing, informatics, and so on.

The concrete objectives of the study programs in EE differ from each other; however, in general they follow the study guidelines that regulate all the study programs within the Science and Engineering Faculty (see Appendix 4).

According to the head of the study board, the PBL concepts were implemented as the study form in these study programs since 1974,[4] which experienced changes along the past 30 years. In general, each semester students carry out a project on a specific topic in the form of group work, except for the

[2]The two colleges are respectively Aalborg Teknikum and Danmarks Ingeniør Akademi.

[3]This is my translation of the name from Danish to English. In the Engineering and Science Faculty, AAU, all the first year students study in the shared buildings, which are in a separate campus from other parts of the university.

[4]Some of the programs were established after that.

final project for the master thesis which can be done individually. The project work comprises half of the total work time. The other half includes lectures, exercises and a field trip.

Staff members who are employed for the teaching tasks in different programs in EE mostly have a background in EE engineering themselves. A considerable percentage of the teaching staff in general received education from the same department as students. Among the ten teaching staff whom I interviewed in my research, 7 (4 male and 3 female) graduated from EE as students. They were all positive at the practice of PBL concepts based on their own experiences in studying and teaching. The other three staff members had worked as teachers before the establishment of the university, which meant that they used to be teachers in the traditional learning environment. The female professor was a strong supporter of the implement of PBL concept into the educational practice. She worked as the pioneer in the educational transformation process 30 years ago and has been participating in the development of the study form since then. However, the two male professors still showed doubt about PBL concepts as educational methods after they had been working in this model for 30 years. One of them held strong belief the traditional lecture based method of teaching would benefit students better in terms of equipping students with technical knowledge.

These four female employees in my study were the only female teaching staff in EE. They experienced being the minority in the male-centred environment as the price of studying and working on engineering both as students and in the workplace. The three of them who graduated from EE at AAU as students had positive experiences of studying engineering in the PBL environment. All of them (except for one who only worked half time in recent years) once worked in the managing areas in the department. Since their graduation, they also initiated and actively participated in different activities helping female students in EE with their study both academically and mentally.

In this study, information from interviews with the teaching staff is mainly used as background knowledge rather than the focus of analysis. Therefore, this section aims to provide a brief introduction to the context of research site in EE from the perspective of educators. In the following sections, I will move to report the research findings of the experiences of students in EE based on the empirical work of my study from both the explorative study from January–April, 2003 and September 2003 to June, 2004.

7.2 Path to EE Engineering

Statistics did not witness dramatic changes of women's participation into the traditional 'hard-core' engineering, EE, under the influence of the PBL study form. Among all the present students[5] (around 400[6]) from the 1st to 10th semester in autumn, 2003, there were altogether 9 female students. I asked all my informants (both male and female) about their life stories in relation to science and technology study in their early life and the influencing factors leading them to the choice of engineering study.

7.2.1 Choice of EE Engineering

The data in this study suggest that there are different paths for men and women to enter into engineering. Each gender enters the profession with a distinct set of orientations and abilities. When asked why they chose engineering education, the two most often-mentioned reasons by men were: they had been tinkers as children; and they had been interested in mechanics, computers and electronics. For women the two most-often-given reasons were: they had been good at and interested in math and science; and they had influences from fathers, boyfriends or other family members who were engineers. These different starting points, in combination with other influencing factors, play an influential role in the decision making and study experiences of both women and men engineering students.

Personal interest and academic background in relation to science and technology

The most impressive differences between male and female engineering students regarding their decision making on career are (see Appendix 3): men became engineers because of a strong orientation towards tinkering and technology; while women were attracted to it because of the interest and skills in mathematics in their earlier schooling.

---

[5] At the 7th–10th semesters all the study programs in EE transferred into international programs. However, in this study only focus on Danish students without discussion on foreign students.

[6] This approximate figure only covers the Danish students who studied in study programs within EE in 2003.

All the male informants in EE mentioned that as boys, they were fascinated by cars, computers, airplanes, stereos, and radios. They enjoyed taking things apart and putting them back together. This passion of tinkering linked their childhood life closely with machinery or electronics and paved a smooth path for them to technical gymnasium[7] in upper-secondary education and electrical & electronics engineering at higher education. Some of the male informants worked as technicians for a few years before they decided to have higher education to become engineers. For most of them, it is a natural life career for them since childhood.

> "I knew that since I was six years old. I wanted to be an expert of robot, and I am on my way, though there is still a long way to be that..."
>
> (Patrick)

> "I like electronics; I like computers, since I was at the 7th grade, around 13, 14. I would like to study something on computers (at that moment), then I had lots of talks with some students here, before I eventually decided to study electronics engineering..."
>
> (Kasper)

For women, excellence in math and physics (some of them) was a primary factor impelling the eight female informants towards electrical and electronics engineering. When talking about childhood, some of the female students mentioned being Tommy boys; however, none of them referred to tinkering as childhood experiences.

The interest in math in early adolescent period did not drive these female students to technical gymnasium (except for one of them) for upper-secondary education. Seven out of the eight female students chose math gymnasium for two reasons: firstly, they could study math at a high level, and meanwhile they

[7] In Denmark, people need make a choice of 'gymnasium' (upper-secondary education) at the age of about 15. There are normally three choices: language, math, and technology gymnasium. Technical gymnasium is a male-centred place with app. 80% male students, in math gymnasium there is a general gender balance and in language gymnasium there are app. 80% female students. While technical gymnasium is focused on providing access to higher education in science and technology, math gymnasium aims broader and language gymnasium does not provide this access. An overall figure of the educational system in Denmark can be seen in Appendix 1.

could consider the choice for higher education along the way; secondly, they were hesitated by the male-dominated culture in technical gymnasium.

According to them, most of their fellow female students in math gymnasium chose some other education or jobs rather than science and engineering education after their graduation. For them, however, the experiences of studying math at high level during the three years promoted their interest as well as academic competences in science and technology area, which made engineering a possible choice for higher education, despite its male-identified image. For example, Laura mentioned how she felt in love with math and physics in primary school because of the interesting teachers. Those experiences worked as motivation leading to her career choice later.

> "I chose math gymnasium because there were so many boys in technical gymnasium. I did not have many male friends at that moment, so I got nervous about the idea of going to a school where there were only boys. I was afraid of not fitting into and being good enough for what they are doing. When I finished gymnasium, I thought it over, and engineering was the thing I wanted most. I knew that there would be lots of guys in the engineering university, but I still would like to study it, so I chose engineering university."
>
> (Laura)

Laura's story is quite typical regarding women's choice of engineering. Except for Line, who had clear goal of becoming an engineer since childhood (because her father was an engineer), these female engineering students experienced similar drifting process before they made up their mind to study engineering.

Influence from family

Among all the informants in EE, about half of them had one or more family members who had either education or profession as hard-core engineers. Nevertheless, this did not play any special role in the career choice making of the male engineering students. For them, the chief and direct reason of tinkering orientation and capability of technology made their decision making process natural and simple. Nevertheless, this research finds that family support can influence women's path to the pursuit of non-traditional career.

All of the women engineering students in EE spoke of the support they achieved from family. In the current Danish society, they mentioned, young people are expected to make decisions on their career choice based on their own interest and will. However, most of them also said that, the respect from family on their intention of continuing the study of science and technology at a higher level as well as the choice of engineering as a future career served as a strong energy resource for their decision making on this non-traditional profession. In some of the cases, their parents supported them to do what they were interested in no matter whether it was regarded as non-traditional profession for women or even they had little knowledge about the profession themselves. For those who came from professional family, they had influence and support from their father, sister, or boyfriend who are engineers, this background worked as both mental support and professional assistance in the academic performance in studying engineering.

Line was introduced to electrical stuff by listening to her father's daily work experiences. The increasingly growing interest in engineering drove her to technical gymnasium for upper-secondary education, as the only female student among all the informants in this study who has this background.

> "My father is an electrical engineer. Since my childhood, I got to know a lot about engineering work from what I heard from my father's daily work life. Later I became more and more interested in engineering profession and wanted to become an engineer when I was at 8th grade in primary school. So I went to technical gymnasium to have more courses and training in science and technology, where I both studied math and physics at high level and got some training in hand-on skills."
>
> (Line)

Line's solid background of technical knowledge from the family influence and support paved her a clear way not only to the later choice of engineering university with specialization of electrical and electronics, but also to a good academic performance as well as professional confidence.

Freda's experiences showed both the supportive and deconstructive influences from private life.

> "I used to work in a company as technical assistant. The job was very focused on drawing. I would like to have more challenges and I felt that a good way to change the situation was to receive more education. My boyfriend was just not happy about my idea because he wanted me to get a job as soon as possible so that I could make money. I decided to choose education instead of that boyfriend because the wish for higher education was so important for me at that moment."
>
> (Freda)

Freda made a choice of engineering education, because it was a more challenging job. This decision surprised her family somehow, not because of the word engineering, but rather the branch of electrical and electronics.

> "They knew that I liked math so much, but they expected me to choose chemistry, biology or something else rather than electrical engineering."
>
> (Freda)

However, her parents gave her warm support to respect her choice. Freda met another boyfriend later at Aalborg University, whose, together with her parents' strong support and encouragement helped her get through the most difficult part of the study life.

Influence from other resources

Sometimes informal advice or encouragement was preferred by enthusiastic teachers might play an influential role in the decision making of some women, like what happened to Laura,

> "When I was small, I wanted to study math, because I had very good math and physics teacher when I was in primarily school. I developed my potential because of their simulation and supports."
>
> (Laura)

From the interviews, women engineering students also mentioned different other reasons that helped them make the final decisions, for example, discussion with friends, teachers in gymnasium, study counsellors, and people around who knew something about the engineering profession. This was something that was seldom necessary for male students.

7.2.2 Choice of AAU

Once having decided to take higher education in the field of engineering, the students need to choose an educational institution. They experienced a process of researching, comparing and making a choice. With a clear ambition of achieving higher level of education, they were facing two options to choose: The Technological University of Denmark (DTU), which is located at Zealand, Copenhagen area, and where a traditional lecture-based learning environment is provided; and Aalborg University (AAU), which is located in North Jutland, and where a student-centred learning environment is provided. Situated in North Jutland, Aalborg University has mostly attracted students from North, West and South Jutland. This factor goes with most of the informants in this study (see Appendix 3). However, when asked about the reasons why they chose this university rather than other options, the most cited reason from these students was the study form. However, there was different concern for the PBL study form.

For the women students, the decision of AAU was based on a thorough comparison between the two different study forms in DTU and Aalborg as well as a careful consideration of their own characteristics and preferences in learning.

Juli has clear awareness of her own strength, weakness and preference regarding learning. She did a research on different study forms before making a decision. And she had a clear goal in concern with what she would like to learn for the self development.

> "I thought a lot about it before I made my decision. I wanted to achieve a master degree, so there are two options. One is at DTU, where the way of studying is to go to lectures, like the way in gymnasium. We had the teacher standing there, telling you something, then afterwards you doing reading yourself, and solve something by yourself, you do everything by yourself. That's the way I was used to and I'm best at, the way I normally learn best. What we do up here (at AAU)

> is that we work in our own study group, and do a project together. And that's something that I was not good at. So actually, I chose to go to Aalborg to study so that I would improve my skills in teamwork."
>
> (Juli)

Similar with what Juli said, the feature of group work in this study form worked as another attractive part for these female students. For them, it can bring comforts and hope when they were confronting dilemma: on one hand they have interests as well as academic skills; on the other hand they kept the insecurity and inconfidence in the study of technical contents. As Mia said,

> "I knew that I was not good at motivating myself, so I need a group to work in, not for depending on the group members to do the job for me, but for myself to be responsible for work. It was also a way of motivation for me. I need that. If I have to do all the technical things alone, I would not be able to finish..."
>
> (Mia)

Laura shared the same consideration regarding group work as a possible strategy of learning technology.

> "Once in a camp, I got a task that is to write a paper in a week. I wrote something about electronics and physics. It turned out to be really really bad because I could not finish the assignment on my own. There were three days when I could not do anything, because I could not make it further. Then I knew that I could not do it on my own even though I liked it, and doing it in a group make me feel more confident."
>
> (Laura)

Sara's story combined different reasons mentioned by most of the women students: to take challenges in life, to make decisions on changes, to pursue personal development based on interest and awareness of the self.

> "I grow up in North Jylland. I would like to know new people and new things, so I chose to start my education in Copenhagen (DTU). But I did not like the way of studying there.

> There were about 200 students in one lecture. We had small projects once a year or once each second year. We have some compulsory courses and some projects based on some certain subjects, but not each year. I don't know about others, but I just could not learn by only sitting here and listening to somebody talking. Also, I would like to have some social activities rather than only school study. That is something I did not have in DTU by working mostly individually. Here at AAU, working in groups itself involves lots of social activities. Well, I studied one year in DTU before I moved here. It is worth moving even though I wasted one year. When I moved from DTU to Aalborg, I discussed with my boyfriend, but I actually had made the decision at that moment. My family supported me along the way. They knew I would study engineering anyway, because my sister was an electrical engineer as well. She studies in AAU because of the study form..."
>
> (Sara)

Compared with female students, the response to the question on the choice of the university from male students were somewhat different. Technology (to be concrete, electronics and computer, remained the first thing coming into their mind when making a choice about education. Then afterwards was the consideration of study form. Different from female students, whose consideration was mostly focused on group work, these male students mostly mentioned the problem solving and project work part of the PBL study form. Because these features of the study form in AAU made the study process more challenging, exciting and more close to the real engineering work. Especially for those who had a background of technical gymnasium, where a similar study form was provided due to the influence of AAU, it was almost a pre-designed career path for them to move from technical gymnasium to some engineering departments at AAU.

7.2.3 Pre-knowledge about Engineering Profession

When asked about the image of engineers or engineering work in their mind before they made the choice, the most cited words men students used were 'problem-solving', 'technology-related' and 'interesting'. In contrast, the most cited words by women were 'hardworking', 'geek' and 'boring'. These women had clear awareness of their own talent in science subjects as well as strong wish to apply it to 'something'. However, most of them (except for Line) had no clear clue regarding how to utilize the scientific knowledge. As Laura said,

> "I love everything about electronics, I love it. I wanted to become an engineer, though I did not know what it was."
>
> (Laura)

They had little knowledge and understanding of what engineers did in daily practice, though some of them had fathers who worked as engineers.

Therefore, this study witnessed a different path to engineering for male and female students in EE. There has been a clear determination for men to enter engineering profession based on tinkering passion, dream to be become engineers since childhood or early adolescent years, experience of machinery, wish to work with technology and suitable gender role of having engineering as future occupations. However, there was a striking sense of drift in the process by which these women chose EE engineering. Having little hand-on skills or experiences in machinery and no orientation towards technology in childhood, most of the women had not thought about engineering until some time in the two years before they left school and went on to university. They were determined to pursue higher education based on their capability in science subjects and good performance in upper-secondary level, but they had not necessarily decided on a future in engineering. Most of them drifted into engineering because they wanted to make use of their own talent and utilize the scientific knowledge they had obtained.

7.3 Being Newcomers

Having been enrolled as students in the programs in EE, these students started their 5-year-long journey in engineering education. Data from the program of EE indicated visible differences between men and women students in general, and among men and among women individually, with respect to getting used to the first year life as an engineering student.

Most of these female students talked about their expectation and feelings about the gender distribution on the entry of this program. Expect for Line, who was not shocked being one of the few female students at the beginning because she had been used to the situation from the past experiences in technical gymnasium, all the other women experienced a period of being surprised and feeling strange by being the extreme minority in the male-centred environment.

> "I knew that there would be lots of men, but I did not know that there were only so few girls. I expected that there would be around 5–8 girls in total. The first day was a gathering of all the students; we were given some introduction before we formed into groups. There were altogether about 80 people. I was so surprised and kind of nervous when I found that there were only the two of us."
>
> (Laura)

Being extreme minority among the male peers brought out a feeling of insecurity to these women, (even though sometimes the only female students were appointed to the same group with the purpose of avoiding them feeling alone). Studying in an environment where the majority of teaching staff and students are men, working in the project group together with five or six men, having no preparation for the high demands in perspective of technical background, these factors led to inconfidence for these women as newcomers of engineering.

> "I did know some of the requirement to be an engineer, especially the math part. But I did not expect so high demand for the electrical part; actually I was not so used to it. And the programming, I like it, but it was quite demanding at the beginning. For a long time, sitting in the group, I was not sure what to say in the group discussion. I was afraid to say anything stupid in front of those guys."
>
> (Sara)

Laura was still struggling against the inconfidence from the lack of technical background when she was at the end of the 3rd semester.

> "I have been working with 5 or 6 guys in the group since last semester. We did not start from an equal place. All of them either come from technical gymnasium, or they used work as electrical technician. So they really knew more than me.

> I think that things would be different if I were from technical gymnasium. Then I would have more knowledge that I could contribute to the project work. Now I don't really have it. Actually some of the things we are studying now are something they have learned already in technical school. I think that later we will be equal, so maybe after this semester I will be much better."
>
> (Laura)

Differences in perspective of technical background made the experiences of being newcomers different for men and for women, and also different among men. In general, the male group members whom most of these women students worked with had a stronger technical background either from gymnasium or previous work experiences. Even those male students who had a non-technical gymnasium background seemed to have an easier journey than these women.

> Interviewer: "How is the situation for those guys who don't have a background of technical gymnasium?"
>
> Laura: "I guess that they felt somewhat similar way with me at the beginning, but they got it through quite fast, because some of them are very experienced after working as electrical technicians. And they are guys, they grow up with those things. Some of their fathers are engineers, and they can always ask questions. I didn't know any engineer."
>
> Interviewer: 'Who will you turn to ask when you need help?'
>
> Laura: "Some of my group members. Some of them are not very nice; they really made me feel that I was stupid. However, some are really nice and helpful. I learn much more from them than from the lecturers."

When asked about their experiences in the first year, responses from male students who came from technical gymnasium indicated a natural extension of study in the area of technology to a higher level — similar study form in perspective of problem-solving, projecting doing and group work, just more advanced theories, higher demands of technical skills, more real life related problems, more challenging projects, more serious group work. For those who came from math gymnasium, it did not take them long to get used to this study form because of their strong devotion to technology study and tinkering background from childhood. For those who used to work as technicians, it was not difficult for them to get involved in this study form because it was workplace-imitated. Having no background of technical gymnasium, Marcus felt that it was quite easy to get used to this study form.

> "I think that it is very natural to learn this way. I had a background as a technician, which means that the job was based on solving problems in life, and I always work together with 2 or 3 people. So we have to be able to cooperate. In this way, it (the study form) was not new for me."
>
> (Marcus)

Hard learning contents as well as the insecurity from being the minority in a male-centred environment made some of the female students realize that they were in the wrong place. Most of these women talked about the leaving decision made by their fellow female students.[8]

> "In the first semester, there were actually around 6 girls altogether, one girl in each group, but later some of them dropped out. I became one of the few left. There was a very feminine girl, she could never be thought of as engineer, and she dropped out later. She had a hard time staying there, making herself heard, and making her opinions accepted by the guys. She said that she did not want to be an engineer."
>
> (Sara)

When asked about experiences of working together with female students in the same group, some men talked about how they indirectly 'pushed' the 'leaving' of some female students.

> "Some of us have tried working in the group with girls, but we scared them away. There were two girls in our group at 3rd semester. In two months, one of them dropped out. In

[8] Informally I was told that about one third to half of the enrolled female students would drop out at the end of the first year. However, the official dropout information was kept confidential.

> another month, the other dropped out too. Well, I think that it is ok to say it now. I think that there were two reasons. One, it is too much for them, I mean, basically the technical stuff, mechanical things, which was much harder than that they thought. And then they were in a group where there were four guys, and we were really loud speaking. We guys could get involved into discussion so much that we just talked so loudly. These girls were mainly sitting back, and keeping quiet, and suddenly we would realize that we should ask how they think about things."
>
> (Male group)

When facing difficulties in the first year experience, support and encouragement from family and other resources turned out to be of great help for these women to continue the education.

> "I really had a hard time in the first year. I was so ambitious in getting higher education and I was quite interested in learning these things, but when I eventually came here, I realized that I missed lots of things. Staying alone in a new city facing these difficulties in study, I thought about dropping out and moving to another university that is close to my hometown. Without the support from my family and boyfriend, I probably would not have made it. I also have met some nice teachers here, they encouraged me a lot..."
>
> (Freda)

A visible difference between male and female students regarding technical background when they entered engineering education has been witnessed from the interview data. The gendered experiences in being newcomers identify some hidden expectations in terms of technical skills of the newcomers, which are based on the past experiences of the majority of male students. These expectations bring about special difficulties and inconfidence to female students because they did not have the same starting point as their male peers.

7.4 Learning Culture at EE

This sector reports the learning processes of these students in EE based on data from interviews (with the eight female students from different semesters and the eleven male students from 8th semester) and observation (with one mixed group and two male groups). During the interviews, students talked about their perceptions on learning, and how they experienced the study life in the PBL environment when studying the hard-core engineering. Combined with field notes from observation, different aspects of the study life are depicted, like learning from attending lectures, meeting supervisors, doing group work, participating in social activities, etc.

7.4.1 Students' Perceptions on Learning

In this study, from both interviews and observation, I found that 'learning' was the most important objective when students talked about their experiences of studying engineering. In the daily practice, students very often related the activities to the discussion regarding what way and to what extent they could learn. In the interviews, most students brought about the topic of learning directly when they described the purpose and reasons of doing things in life. For example, when asked about the current project which Line was doing, she said,

> "It is really exciting. It fits my way of learning. I learn better when I find out the way myself. This way of learning is much better than only attending lectures, because I have to know why I need learn this. When I know the objective clearly, I learn much better."
>
> (Line)

Group work[9] has been mostly mentioned as a beneficial way of learning both as a spiritual support and technical strategy. Most students talked about

[9]In this study I do not distinguish 'group work' and 'team work'. However I mostly use the term 'group work' because 1) the direct translation from Danish 'gruppearbejde' to English 'group work'; 2) Wenger (1998) distinguishes 'communities of practice' from 'team work' for the three characteristics (as discussed in Chapter 3).

supports that were received from group members as well as responsibilities that were developed from working in groups. Working in groups was particularly referred to as a supportive way to keep women who had strong wish to study engineering from dropping out.

> "This has been so tough, even though I liked math so much. If it was not because of the group work, I would have dropped out years ago."
>
> (Clara).

Teamwork (the same with group work in this research) is part of the nature of engineering, which is something many students had strong awareness of and had the intention to do. Through doing teamwork, peer learning was appreciated and well-used as a learning strategy by the students because of the efficiency it could bring both in terms of time and information resource.

> Marcus: "We can also be capable of solving tasks individually. But the fact that we are in groups is for the purpose that we can support each other in order to reach the goal of the project. Mutual support is one of the best things of this learning environment in Aalborg University. You may work individually, but when we are working together, we discuss about the issues and share experiences."
>
> Karsper: "for example, if I have a problem, it might take 10 hours to solve the problem. Maybe it is just because I am new in this area, then I can ask the group member, who knows more about this field. Then maybe he can help me to make it clear in some minutes. It can make things so efficient sometimes."
>
> Michael: "sometimes just by discussing the problem with the group members. Maybe I don't know this area either. But if it takes him 10 hours to figure it out alone, it might just take 3 hours when we discuss about it and work on it together. Then both of us learn it."
>
> (Male group)

Six out of the eight interviewed women students mentioned the inefficiency of learning technical stuff by only sitting in the lecture room and listening to the presentation. Group discussions proved to be a helpful way in terms of understanding technical knowledge. Juli, one of the only two women who claimed that they were abstract learners, meaning who could understand the abstract technical theories pretty well just by reading books and listening to lecturers, talked about her appreciation of learning through doing group work.

> "I think that it turns easier when you learn the technical knowledge in groups. Normally we use the blackboard to discuss things. We set up some problems that needed to be solved. We do it in turn. On shift, we go to the board, talking about the solution. Everybody comments on that and talks about the opinions on how we can solve it. We gain more from the time we have to spend at university in this studying when we work in teams. We are getting energy in this way."
>
> (Juli)

Doing group work involves a process of reflection, in which students can on one hand speak out their own mind and get some feedback from group members, on the other hand they listen to others' and think about comments as well as comparison to their own ways of working. This is also an efficient way that helps students have more knowledge about their own way of learning.

> "It is a lot better than what I thought it was, but maybe it was because I was not used to working in groups before. When I worked by myself, I got help from friends, then I knew that it was good. When I work in the group with 6 other people, I can see what they do, that is what I like."
>
> (Laura)

Having strong interest in math and technology, but at the same times showing worries about working in a male environment, Laura found this way of learning in practice better than her expectation.

7.4.2 Learning Processes in the PBL Environment

In Chapter 2, I have discussed the principles of the PBL Aalborg Model, which is characterized as problem-orientation, project work, interdisciplinary, participant (or self) directed learning, and exemplary principle and team work.

These characteristic elements have been examined through the empirical work in order to draw a picture of how these function in practice and how they influence the learning process of students. Based on the theoretical study of learning, PBL concept and engineering education, this study had an assumption that the learning resources for engineering study in the PBL environment mainly come from doing projects in groups, attending lectures, and getting inputs from supervision. The following questions were kept in mind in the course of field work. How do these resources function in practice? In which way do they influence the learning process of students in reality?

Doing projects in groups

Doing project in the form of groups involves lots of things to learn. According to the interview with the head of study board and the description of the study guideline, there are three major learning objectives. First, students are expected to develop themselves as problem solvers. Second, they need to learn how to learn through participation into projects. Third, they are expected to learn to organize and collaborate within a project organization. Students are expected to develop competencies within areas as problem solving, project work, collaboration, and learning to learn.

Problem formulation and group formation are the first activities of each project. At AAU, at the end of every semester, there will be a number of teaching staff appointed, who will be available to be the supervisor in the next semester. Each of them is expected to propose a couple of general problems in a certain research area and publish them on the website so that students can have a brief review. New semesters start with topic choosing and group formation. The first day of each semester will be an introduction activity with all the supervisors and students from the same year. Supervisors will make a public presentation about their proposals for students to make a choice. The coordinator will put all the proposals on the blackboard with different categories. Students can ask questions and have discussions before they put their names into different categories. Normally[10] students are supposed to find their interest in some specific topic, based on which, groups will be formed.

[10]Except that in the first semester, students are appointed to groups in a one-month-long project (P0 project), which aims to provide a chance for trial.

Interest in the topic and the right people to work with are the two main concerns in terms of group formation. In the first two or three semesters, students mainly choose the topic based on their interest and regard working in different groups as an experience gaining process. Most people experience lots of problems arising from working in the groups in which there are people they can not work together with due to different working styles and personalities. When reaching 5th or 6th semester, they feel that they have spent plenty of time and energy on learning how to collaborate and finding out how to make the group work function. Thus, who to work with, instead of what to work on, turns to be the first priority when students form the groups since 5th or 6th semester. According to most of the informants, as long as they found people whom they felt comfortable working with, they could always find interesting topics to work on. Therefore, group formation turns out to be the most important starting point for the project work.

> I: How did you choose the project for this semester?
>
> R: The first day of each semester was introduction day, when we formed the groups. There would be some invitation of topics presented by supervisors. We could also make our own topics. There were 10 topics to choose and every group had to deliver a priority list with three topics. Then the supervisors will sit together and distribute themselves to each group.
>
> I: How did you form this group?
>
> P: we were in the same group last year. Before that, some of us have been working together for two years, some for a couple of semesters. We were quite lazy, so we just keep this group. (they were discussing and joking how many years they have been working together).
>
> P: At the end of last project, we went downtown to have a beer, then we decided to continue to work together.
>
> R: How did we four come together 2 years ago? Oh it was also when we were downtown having a beer, we decided to work together.

> M: I think that in general the group formation is kind of trying-out. Years ago, when we started to have experiences and found out there were some people we did not want to work with any more. Then we heard about each other and we would like to try to work together. While then, there were also some people we did not know about. So we could try to work together for half a year to see how it could be.
>
> P: I think that in the first 2 or 3 semesters, it was mainly coincidence. Then we found that we could manage to work well together with each other after 3 or 4 years experiences together.
>
> (a male group)

After the group formation and the appointment of supervisors, students in groups start to find out what they would like to do in the project. In the practice of PBL, the process of problem formulation is not carried out as what it is ideally described as, that is, a way students probe problems themselves. Rather, in reality, it is often that students define a small and concrete problem under the umbrella of a big problem proposed by supervisors. Nevertheless, it works as a strong driving force for students to initiate learning. As can be cited from most of the students interviewed, 'when working on a problem, I am strongly motivated and attracted, because we need to solve this problem …'

The majority of the interviewed students saw the achievement of visible capability from themselves through solving problems. Most of the women students said that when compared with their female friends, they realized the improvement in themselves. As Sara said, 'I can feel the difference when I stay with my friends who are outside of the university. I can feel that I am using different ways of solving things. But it came naturally.'

These students also mentioned different bad experiences during the first year: difficulties in understanding the goal of learning, conflicts in doing group work, different expectations in technical knowledge based on different backgrounds. After a couple of semesters, they started to realize that the main objective of learning in PBL environment was to learn how to learn.

> "In the first semesters, we made lots of mistakes, then we won't forget those parts. In next project, we will remember

> the mistakes we have made .... We need remind each other that 'Hej, here is the place we once made mistakes in last project, we need to be careful' .... Then we discuss about it, why it was a mistake before, how it happened and whether it will be another mistake in another case, and how it can be done in another way, and so on .... We really learn from lots of discussion like this ... I think that this is the idea of this study, we learn from our own mistakes, and then we learn how to learn."
>
> (Discussion in one male group)

When asked about what are the most important factors that make the group work function, positive attitude for cooperation and agreement came out to be the most cited replies from both male and female students.

> "We have very positive attitude to cooperate with each other. In this way we have the strong belief that no matter how different we are, what kind of disagreements we have, we will manage. That is the most important thing that motivates us and helps us go through these years."
>
> (Clara, Christian and Karl)

Just as most of the interviewed students agreed that a good attitude to collaborate was a fundamental element to support the whole process of doing project in group work, they also referred to achieving agreement as the most essential thing to get the group work move on.

Many students reflected on what they have learned in a course, Collaborative Learning in Projects (CLP), which was provided at the first semester with an aim of teaching students how to learn through doing group work.

> "There were some courses teaching us how to communicate in the groups in order to get agreement. I think that the courses helped a lot, especially at the beginning ... It was not because some people had problems but rather, but just because we came from different gymnasium, and we were used to different ways of working ... Some people were more used to work alone, and in this case, they would like to do things in

> the way they wanted to do things individually, but working as engineers involves teamwork. It is the way many engineers do... So we need learn how to cooperate and work together..."
>
> (Male group)

Making a collaboration agreement was a strategy they were taught in CLP course during the first year. By doing that, they established some agreed values and norms while working in the group. It helped them to go through the first period of student life in perspective of collaboration with people who had different personalities and working styles. When they became more experienced in later years, this collaboration agreement turned into hidden-form.

> "We don't have collaboration agreement any more. We used it in the first two semesters. The most important thing is to agree on things, and then we don't need have that written down. We know each other in the group and we trust each other."
>
> (Mia)

From the written form of collaboration agreement to hidden agreement, it is also a process of developing responsibility as well as tacit knowledge.

> "When we were younger, we wrote it down in a paper. Now we are more mature, we know the importance of being responsible for ourselves and for the group members, also we have the trust. We know that it is serious work, it is for the learning of ourselves."
>
> (Male group)

To get a project done efficiently by 5 to 6 persons involved dividing the work into different tasks. Therefore, doing group work in groups incorporated doing individual work, working in sub-groups and working in the project group. Many groups put both long-term (normally a semester, for the whole project) and short-term plans (either one week or two weeks) on the wall of their group room. For the long term plan, they signify some milestones on a semester calendar, which might be kept flexible for modification along the way. For the

Picture 7.1 project planning.

short term plan, some groups draw time tables on a blackboard (it is normally the task of the group leader), and in the group meeting they fill in things that need to be done before they discuss who will do them. Some groups make a list of tasks in the group meeting and then discuss who would like to do what and how long it will take.

When asked how they distribute the tasks into sub-group work and individual work, most of the interviewed students talked about the principle of 'making sure that everybody can learn as much as possible'.

> "It will be based on individual interest. Normally I pick up things I really want to learn, sometimes it is because I am so attracted and eager to learn, sometimes it is because I know that it is something I am not good at, and I would like to improve."
>
> (Maja)

When students gained more experiences through working in groups for a few semesters, they had more awareness of the advantage of peer learning. At the beginning of each project, they would talk about their own strengths, weaknesses and expectation of the new project. In this way, they distributed group tasks with strong awareness and a clear goal.

> "We talked about many things from the beginning so that we knew what each other was good at, and what we could learn from each other and what we could like to learn through doing the project. We want to make it possible that we can learn from each other and everybody can learn what they want to learn."
>
> (a male group)

This awareness was shared by most of the groups from the interviews; however, this was not the case in the first year, in the daily practice, especially in the first two years, sometimes priority will be given to temptation to make fun and doing something comfortable. For example, laboratory work is a task to practice hand-one skills, similar to what has been reported in different literature, it remains one of the most exciting things for male students and non-attractive work for some female students. A couple of female students said that they found it interesting when they were supported and assisted by their male group members. Some female students talked about their unconscious decisions to escape from it and let their male group members do it. For those women who claim 'I don't like it', the reasons can be traced back to the inconfidence arising from the lack of technical background. As Sara said,

> "I hate that part. I did not like it because I was not good at it. I did not know how to do it when I have to put these little thing together with that... i don't know what to do. Yes, I have the knowledge but I never think about how to apply it.... After two semesters, I just still did not know how to do it. I never did it the right way. Boys love that, because some of them think writing report is really boring. I realize that it is kind of stupid thing that I did not go there (lab), otherwise I could have learnt that."
>
> (Sara)

This was the experiences in the first two years. When Sara was at the 8th semester and looked back to the past experiences, she reflected that she would have been able to do the laboratory work and learned more things if she could just be brave enough to try it and do what she was supposed to do from the beginning.

To select a group leader was another strategy they learned in the first year to make the group work function. The group leader was expected to take the responsibility of the group organization, whose concrete tasks should be decided when group members made the collaboration agreement. In some groups students took turn to work as the group leader, and in some groups some students volunteered to do the job. Most students regarded it as a useful and efficient way to improve the capability of management as well as sense of responsibility. Most people had the awareness that 'if I am going to work in the companies in the future, there will be chances to work as project manager, so this is a good chance to learn.'

Working as a group leader had been good experiences of most of the students in their first years' education, though they stopped using this strategy in later years when they became more experienced in reaching agreement and managing learning. According to male students, the experience of working as group leaders helped them practice organizing things as well as build up a sense of responsibility. However, it worked in a different ways for female students. A couple of female students said that they never had this experience, because it involved being different and it might bring uncomfortableness when working among all the guys.

> "I was never that type. So I was always to be pushed to things... They have been complaining of me for not being leader or being that way. I never like that, because I think that it was a waste of time to compete... I like the situation when there are no visible leader types, so we are equal and we can get relaxed together."
>
> (Laura)

Nevertheless, some other had different attitudes towards and experiences from working as group leader as a resource of confidence.

> "I haven't been that (group leader) before in the past years, since being the only girl, you won't want to do it because no people will listen to you. When we started to have two girls

> (in the same group), I tried it. It was really good, I am more confident, at least I know that there will be somebody who will listen to me. When I said something as a leader, they (the guys) listened to me, when they listen to me more, I will get more confident. It is a circle thing."
>
> (Sara)

In the process of doing group work, different expectation and ways of perceiving the education can be witnessed from male and female students. For male students, it was a career they were being trained for and this education was leading to a profession. For female students, it was mainly an educational program they were participated into in order to learn the subjects that they were interested in.

> "I think that when lots of people study here, they want to play as engineers; they take things seriously like they were in real companies. I don't really want to be that. I want to do what we have to do, and then have fun, and not to be so serious all the time."
>
> (Laura)

Working in groups provides chances to getting to know the strengths of each other and learning from each other. It is also a process of developing one's own potentials and bring different contributions to the shared work. When asked about their own strengths and contributions, male students normally would mention different technical areas that they were specifically proud of themselves. Women students would refer to drawing, calculation and language edition in writing as their main technical contribution. In addition, women students all spoke of their special contribution in terms of planning, coordination and management. According to them, male students talked a lot and they could get very involved in discussion, but they were not good at keeping limitations and documenting. They tended to forget deadline and things that had been discussed. Therefore, female students often took the responsibility to make sure that all the meetings would be documented. Most of the time in the group work they played a role of time keeper and program organizer. As Juli said,

> "I am good at structure. I am the one who is arranging things, making programs and agendas for everyday, structuring what

we are doing, and so on. Guys are messy, sometimes they just want to make fun and tend to forget deadline. They will always say that 'don't worry, we will find out..."

(Juli)

Most of female students (except for two) mentioned programming as the most difficult part for them, particularly in the first year. According to them, it was mainly because they lack sufficient background knowledge, and programming was publicly regarded as 'guys' thing. In order to make it up, some of them took some extra training program in their spare-time. Or sometimes they could get some supports and help if they were lucky to have some nice and patient male group member. Sometimes by asking questions to male members, it brought about a learning process itself. As Laura talked about her story,

> "I don't have so much knowledge as they do, because of the gymnasium background. I have never done so much about electronics before. So I ask them lots of questions like 'what do you mean by this or that?'. That helped them understand things better, because when they explained things to me, they could understand more."
>
> (Laura)

Sometimes women's value and contribution could be appreciated by their group member, as Christian and Carl did. They said that it had been a very good experience working with Clara these years. They learned a lot from her, because she was academically capable, intelligent, structured, and well-organized. As they said,

> "We always got good marks because of her, and it has been enjoyable working together with her. She is as logical as guys, she is precise and strong-minded. She took our humour, and she thinks different from us sometimes, that makes it quite fun. You feel something different."
>
> (Christian and Carl)

However, many of the interviewed male students had no experiences of working together with female students. They were used to working homogeneously and they did not think that engineering was the right place for women. Some of them had some 'bad' experiences of working with female students because 'they did not know what they were saying in the group work.' These male students shared one common idea that women did not match engineering and only those who were 'intelligent enough' and who 'knew what they were doing' could be welcome.

In general, both male and female students reflected on the learning process they had experienced as a development in which they grew up and became more mature.

> "I learned to handle those problems gradually. I am more and more used to different situations now. And I also know more about myself, to be aware of what I am good at in the group and contribute to the group work. Working in different groups have been really challenging in life. But we grow up along the way; we know that we have to find out a way to get the group work done. I felt that I have experienced a lot more than my female friends who are doing something else or who take some other education. Working in groups really helped a lot in this way."
>
> (Freda)

Having experienced plenty of conflicts through doing group work in the first years, Freda learned to develop different strategies to get the best out of different people she worked with. She also recognized the maturity that had been developed along the way.

In general, the research process observed a diversity of individual features in doing project work, especially among male students. Nevertheless, an overall gender pattern can be recognized from women's strengths. These female students are in general better at managing, planning, organizing, coordinating communication, and so on. Male students in general appreciated the communicational skills that had been developed through group work, however, for them, these skills were not counted as engineering skills. Technical skills are still regarded as the most important contents for the engineering study.

Lectures

In a traditional learning environment, lectures remain the major resource for learning, which occupy over 80% of the work hours of students. Students read literature, attend lectures, make notes and accomplish assignments afterwards. Different from this, lectures occupy about 50% of the work hours of students in PBL environment at Aalborg University. Nevertheless, lectures do play a role in the learning process of students. The objectives of providing lectures seemed to match the expectation from students in general, that is, to provide (teaching) and to gain (learning) new knowledge as inputs in some specific fields. There were no obvious gendered patterns in terms of the reflection on the function of lectures. According to the interviewed students, 'we are basically different from each other as individual persons'. During lectures, students were introduced theories. To what level they could understand these theories depended on different factors such as teaching methods of the lecturers, cognitive learning styles, interest and background knowledge about the subjects, schedule design of the lectures and conditions the learners were in, and so on.

In general, there was some common evidence regarding in what ways lectures would lead to better understanding and more efficient learning. There were generally two kinds of courses provided every semester, SE courses, which provide background knowledge in some subjects and will be assessed in the form of written exam, PE courses, which provide knowledge that can be directly related to project work and will be evaluated through project report and the defence (oral exams).

All the interviewed students talked about the important role of application and practice play in terms of learning from lectures. In general, the students felt that they would have a deeper understanding of theoretical concepts when the teaching contents were reflected through doing projects.

> "I think that it is quite normal to forget about things you take in from the courses if you don't have chance to practice and use it. What we can get most out of the courses are the things that can be used in the project."
>
> (Simon)

Both PE and SE courses[11] included two hours lectures and two hours exercise time. This way of arranging courses was greatly appreciated by all the students, because it provides a natural learning process, in which students were taught theories and afterwards were provided a context or some examples to practice through doing course exercises. In this way, students with both abstract-preferred and practice-oriented learning styles would get a chance to learn.

> "Lecturers give you kind of indications — what are we talking about here? What is the motivation? What is the subject? But normally you don't really understand it during the lectures. Then you come back to the group room and sit down, facing the problem. You are motivated when you get into the thinking process when trying to solve the problem in the course exercises, especially when they are related to the project. You can look at the computer and get some experiences by doing it. Then the lecturers will visit the group room, and you can talk with them, who can explain to you things in a situation and context. In this way you can gain more understanding. That is how the Aalborg model works, I think."
>
> (Victor)

The expectation of lecturers fell into two aspects: technical knowledge and the art of teaching. Technical knowledge is a rather important way of getting direct inputs; however, a very knowledgeable professor and capable engineer was not necessary a good lecturer. The students experienced different lecturers who were very skilful in their own research and knew a lot about the field but had difficulties in passing the knowledge on to the students. Therefore, for students, to have rich professional knowledge was not good enough to help others learn, and it was more important when it was combined with the art of teaching.

Talking about their expectations for the art of giving lectures, the most cited replies cover three aspects. Firstly, it was essentially important for the

[11] In the PBL AAU model, there are two different courses provided: PE courses, that is, project related courses; SE courses, subject related courses.

lecturer to provide an interesting, understandable and informative presentation that covers clear overviews and well-organized structures.

> "I need know the structure of the presentation. Firstly, what is the overview and then what will happen step by step. It will be nice if we can always have the slides beforehand to have some preview. Slides with colours, diagrams that are well-organized can be very catching."
>
> (Clara)

Secondly, it would be helpful for understanding if the lecture can explain clearly the goals and purposes of learning different theories or formulas. They needed get the point directly before they were directed to move on along the lecture.

> "It is so important to use a few minutes to let us know the goal of these formulas. As students, we are very practical, we need see the need and purpose of things in this lecture. If we can't use it for anything, then we can't learn it."
>
> (Patrick)

Thirdly, all of the students emphasized that using examples would be very efficient way to help them understand sophisticated theories. They needed to see the relations between the abstract and real life in order to learn, and examples were good ways to help them get some ideas and see the links.

> "When we are taught some formulas, the theories behind the formulas are very complex, it is difficult to get the point only seeing the formulas and hear the explanations. But if we can see some links showing how the formulas can be used, then it makes sense. Then we can know in what contexts the formulas can be used."
>
> (Male group)

Both the interview and observation data show the students' opinions on the quality of good lecturers, as the vocabularies they used: 'well-organized', 'structured', 'humorous', 'provocative', 'good at explaining things'. Some students talked about the importance of the personality of lecturers. A positive and motivating lecture could significantly influence the atmosphere of the lecture room and improve the effect of learning.

Some subjects in engineering curriculum are sophisticated and difficult, which are regarded as 'killer courses'. In this case, a good lecture will be of significant help. Some hard-core subjects, for example, programming and electronics, can be more difficult and threatening for women students because of the lack of pre-university training. Some lecturers tended to assume that all the students were at the same level. However, a kind lecturer who could keep this difference in mind would be very supportive and motivating in their learning.

Most of the interviewed students in EE had not experienced any female lecturers due to the fact that the majority of teaching staff in this department were male. All of the students said that they did not mind the gender of the lectures. In fact, male students were not positive about the idea of having female teaching staff, because the only experiences they had of female lectures were not very good. A few women students referred to some cases of good female lecturers in the first years. They might come from some other areas, such as designing and drawing, rather than hard-core engineering. The reasons these female lectures left a good impression were not because of their gender, but rather, some good points as lecturers in general. For example, they were well-organized, and they could explain things clearly so as to make it easier for students to understand.

Supervision

In the PBL environment at Aalborg University, supervision plays a role of facilitating learning instead of merely transferring knowledge. Different aspects were referred to about the expectation to supervision. The general expectations for supervision are reflected upon (1) skills of supervision. Most interviewed students expected supervisors to instruct them to learn to learn, that is, to learn the methods, rather than to be provided direct answers. Learning by doing and reflecting was the motto for many students. They enjoyed the learning process of trying-out in practice after being introduced to theories. Therefore, they preferred to learn the tools so as to be able to do the project independently. They expected to be trained to handle things on their own, and at the same time to be reminded of the mistakes and big holes by supervisors before exams. (2) Supervisors' technical knowledge, engagement, and accessibility. The devotion of supervisors could bring about motivation and spiritual supports.

The way supervision function as well as the expectation of supervisors is not fixed, but rather, changing along the proceeding of the project. At the beginning of the whole education and each project: firstly, patience and willingness to help that could make students feel comfortable asking 'stupid' questions. Being experienced, some professors tended to take it for granted that some things were so 'easy' that students should have known, but forgot the fact that they were at very different levels regarding technical knowledge due to different backgrounds. Secondly, inputs on resources of information. For example, to provide possibly useful literature, sources of database, methods of analyzing relevant information, etc. Mia talked about a good supervisor in her mind when she was at the second semester.

> "He is a great guy. He will make sure that we know it. He was always there if you need him. If we had more questions, we could just go to ask for help. Even though sometimes we just expected to have a 'yes' or 'no' question, he would say 'oh, let's talk about it'. Then he could spend one or two hours discussing about it. If there are something he does not know, he would try his best to find it out, either go to ask other people, or check some database, or whatever ways. He would come back to us with something. That was really nice."
>
> (Mia)

Working through project work in groups was also a process to gain experiences in terms of how to be active and get best of the supervisors. As newcomers, students did not know what they could expect from supervisors and how to communicate with their supervisors in order to get help. After the experiences of doing a couple of projects, they learned to express their expectation to supervisors at the beginning of the semester regarding what they intended to learn through the project and what help they expected to get from supervision. They learned to communicate with the supervisor in terms of how to ask for help, how to make positive complaint, and how to talk about problems when there are mismatch. Clara talked about her experiences,

> "In the first year, we had no idea what we can expect from supervisors and how to communicate with them. So lots of problems came out along the way. Now we know what we

> can do from our side. If we have specific problem we have to solve, we will tell him and be clear about what specific help we expect to get."
>
> (Clara)

Similar discussions about supervision were mentioned by most of the other interviewed as well. In general, when students were at the first a couple of months of each project or the first two years of the overall study, they needed direct and concrete instructions as inputs. This need would be decreased when they became more experienced and independent at the later part of the project or education. Instead, it will shift into expectation on structural comments. As reflected by one group,

> "At the first semesters, we were quite dependent on supervisors, but we are more independent now. It changes a lot since the beginning."
>
> (a male group)

Speaking of their expectation for supervisors at the current project, this group mentioned that having experienced 7–8 projects, they were more confident in dealing with different situations. They expected themselves to have the good attitude facing different problems. Therefore, supervisors at this moment should be somebody they turn to for some comments on overall structure and some emergency help when they had problems that they could not solve at all.

Few of the interviewed students had female supervisors. Having no experience, they could not imagine what it would be to have a female supervisor, especially for the male students, because it was natural for them to live in the male environment with only male teaching staff. Nevertheless, all of them said that they would not mind the gender of supervisors.

One of the male groups also referred to a female assistant supervisor in the first year. She showed special concern and asked them different questions on non-technical aspects of the project, for example, the societal application, communication, and group dynamic. However, her work was not taken seriously by these male students,

> "We guess she had a hard job at the moment, because we could not get her point. It was not because she was a woman.

> But rather, we were interested in learning electronic stuff and she was keeping on telling us to look at the human aspect of electronics. Well, we just wanted to make a project and we don't care about the human problem."
>
> (male group)

Helping students improve their social skills like communication and collaboration in the group work is supposed to be the task of the main supervisor. However, in reality, more interests are put in the technical part by both the male supervisors (mostly male) and the male students.

Sara, Juli and Laura talked about the only female supervisor they experienced in the first year. Sharing the same paths to the engineering world, they felt close to her. Different from other supervisors, she provided lots of help in terms of structural planning, making management, and improving communication in order to achieve better learning, etc. They benefited from these help in the following years.

Social activities

Throughout the 5-year-long engineering education, students are expected to work on 10 different projects in different groups. The learning process involves both professional knowledge study and social activities. By working together in the groups, students spend most of their work time together, and in the last period of the project work, they even spend most of the everyday time together. Throughout the years, they conducted many activities together: attending lectures, doing exercises, meeting supervisors, working in labs, having group meetings, writing reports and taking oral exams. In between these formal activities, they also spend plenty of informal moments together: coffee breaks during the work day and lunch time provided chances to chat a lot about different things and to know more about each others, attending parties and going out in spare time were the moments they had relaxation together and saw different sides of each other when being out of work. Through participating in these activities together, they got chances to know more about each other and learning from each other. Sometimes just being together involves social activities, which plays an important role in the learning process.

> "We can learn the professional knowledge no matter by doing group work or individually, however, the social skills in the group work will improve the learning process. It will be more fruitful with better social activities."
>
> (Male group)

In general, all the interviewed students talked about their appreciation of learning through informal and social activities. The most often-heard reflection was the good atmosphere it brought about in which students could enjoy the learning process. Both male and female students mentioned how it help achieve better understanding of the technical stuff and get inspiration of new ideas by discussion with group members. As one of the male groups said that they got new ideas about their project topic for the following semester when they had beer together.

> "Social activities will promote discussions in project work. Maybe inspiration will be achieved through having beer together."
>
> (Michael)

As reported in previous text that the interviewed female students in general regarded group work as a supportive way to study engineering. For them, a good relationship that developed among the group members through different informal activities played an important role in their learning. To develop a good friendship among the group members would directly benefit the project. Especially when they reached the later years of education, life became more university related, because they tend to spend more and more time on doing project due to ambition and interest. In this way, they also started to have less contact with friends outside of school when they reach higher level of the education, since 'it gets more difficult to explain to the friends what we are doing here,' as some students said.

Therefore, it became more important to have a good atmosphere in the project group so that they could get through the difficulties together. However, this friendship was something developed with the growth of maturity and experiences. As Line said,

> "It could be hard to develop and keep this kind of friendship by seeing each other everyday. There were so many problems

coming out. However, after such long time and intense contact, we know each other so well so that we can get along together this way. This make the boring life interesting, especially when the exam time approaches and all we do everyday is to sit in the room around 15 hours a day, being stressed and exhausted."

(Line)

These informal activities also played a role on building up the shared values in the group, which defined what kinds of practice that would be accepted in this group. This, in turn, helped the group members develop a sense of belonging to the group.

7.4.3 Daily Practice of Engineering Study Life — A Day with Clara's Group

This section provides a one-day experience in the learning processes when studying engineering in EE in a PBL environment. This aims to (1) provide a portray of daily practice when studying engineering in EE; (2) confirm and compare with, and make up information to the information provided in the previous text, which is based on interviews data. The main reasons I chose Clara's group were: (1) continual research after the explorative study[12]; (2) consideration that as senior students, they might had rich experiences to reflect on; (3) hope to build up a case of role-model based on Clara's experiences because being the only senior female students she went through the transformation process and managed the engineering study successfully.

Background of Clara:

Clara had good performance in math and physics in primary school. The adolescent ambition of her was to become an engineer, like her father, who worked as an electrical engineer. This wish was mostly based on the worship of her father, though she did not really know a lot about what engineers did. However, an early marriage put a pause to her future education. She had different experiences of traditional women's jobs, like the secretary of a doctor,

[12] In my explorative study (see chapter 6), they were the first study group I observed and interviewed, and Clara was one of the first female students I interviewed with in EE.

as her mother, saleswoman in glasses store, purchasing manager, director of a clothing department in a shopping centre, and so on. After divorce, the insecure finance situation drove her to find a way to make a good living so that she could take good care of her children. She did not want to take any jobs like she did before because they were all low-paid. When she got to know a boyfriend, who she now lives with, she was inspired by his interest in and devotion to electronic engineering. With his support, she made a decision of receiving higher education in electrical and electronics program four years ago.

Clara started from mechanical engineering, but it did not take long before she found it boring. She then changed to electrical engineering with the intention to specialize in medical engineering. She was more interested in the medical part, hoping to combine engineering with health, then she transferred to EE and stayed there for 3 semesters before she felt that she was not good enough for this education. At the 6th semester she chose signal processing as it could make it possible for her later to take a master in medical engineering. Signal processing turned out to be less interesting than she thought it would be, because she found that it was the hardest part of engineering and it demands a nerdy way of doing it. Afterwards, she applied for a transfer to the electrical energy sector where she finally found her self at home. She was very delighted that she made this decision to shift, as she said, 'It is more physical, something you can touch, it is easier to communicate with it. I can relate it to life. It is quite different from signal processing. I don't want to spend my life in front of a computer, analyzing signals.'

The first year has been so stressful that influenced her private life with her boyfriend and children. She was very stressed from the hard work load, new working area and the nerve-racking social atmosphere in the group — being one of the two girls among around 80 students. However, she has managed to get through different difficulties, and now things are getting much better. There is almost no more negative influence, for they have got accustomed to this way of life. However, by working as an engineering student, she won't have more time to stay with her children, which means that they have to be more independent.

At the moment, she was an engineering student in the program of electrical and energy engineering. She lived together with her boyfriend and two children (one daughter at the age of 14 and one son at the age of 8). She took student loan from government in order to get through this education financially, and now she worked hard in order to finish the education and find a well-paid job.

Currently (in November, 2004), she was doing the final project which will lead to the Master thesis. This project is about high voltage protection of large power transformers in cooperation with a Danish power plant. Working together with 3 other guys, Christian, Carl and Jens, they got this topic from their supervisor. It is something they were interested in and have not gone in depth yet. It is a real problem in life, and it would be a good Master project.

An average working day:

8:00 on a Monday morning, I arrived at Clara's project room as we agreed. She hadn't arrived yet. Her group members, Christian, Carl and Jens were making coffee and talking about their experiences on Sunday. I asked them whether I should go to the lecture room to wait for her, and I was invited to join their morning coffee in the project room. They told me that Clara would come in a few minutes, because she was supposed to send her children to school first that day. As a habit, all the group members would meet in the project room every morning before they went to lecture room together. They invited me to join their morning coffee and continued with their chat. During that period when I was following them, they shifted their discussion into English as long as I was there. Christian talked about how he made some progress in changing his website. Carl told us his experiences in the fitness centre yesterday. Jens was standing there, listening to their discussion while sorting out books and papers.

It was a room with about 10 square metres. It was divided into two parts by the bookshelf and fridge. Jens had his desk and laptop in the smaller part which was close to the door. The inside part of the room was the workplace of Christian, Carl and Clara. There put three desks that were combined together, with three desktops on them and four wheeled-chairs around. Besides computers, the desks were covered with different papers and books. In the shelf, the two top levels were filled with books, folders and papers and two lower levels with food storage — packages of bread, fruits, soft drinks and some empty bottles of beer. Besides the shelf stood a small fridge with a small microwave oven on the top and a small desk with a coffee machine and water boiler. On one side of the wall, there was a big poster with a table for long-term plan for the whole year project (they were working on a long project for their master thesis, which would last both 9th and 10th semester). Next to it was a white board with a table for short plans: normally they put two-week schedules on it. On the other side of the wall, there hang a blackboard with different formula and messages and so on. Next to it were different diagrams, maps, photos and commercial pictures, etc.

Previously, they had explained to me that they had all agreed to work on this project because of their common interests and earlier experiences. However, they found out that it was hard for the four of them to work together all the time due to their different ways and preferences of working. As an agreement, they decided to work in a 3 + 1 form, which meant that Jens, who preferred individual tasks, worked on some part of the project on his own. Clara, Christian, and Carl shared similar working style, so they preferred to sit close to each other and work together most of the time. They all enjoyed learning from discussions, from which they could stimulate and inspire each other.

8:05, Clara arrived at the project room. She greeted everybody and gave me a hug. She told me that she got up at 6:30. The first thing to do everyday was to wake the children up and prepare breakfast and lunch pack for them. Some of the weekdays, she drove the children to school on her way to the university (sometimes the children would go to school on their own by bike), and her boyfriend went to university by bike. Christian gave her a mug of coffee and told me that Clara was a superwoman because she could manage to do many different things everyday.

8:13, they walked to the lecture room together. When walking through the building, I felt walking in a world of machines; huge machines could be seen from different heavy doors. The lecturer, Rasmus, had arrived and was preparing the facilities for PowerPoint presentation. They greeted to the lecture and other students and got seated next to each other. There were around 25 students in the classroom. I sat next to Clara, and we were the only two female present. I gained a few seconds' attention from the guys, but nobody really asked who I was or talked to me. I had introduced my project and intention to Rasmus beforehand, and he said that I could just show up because everybody in the university could join any public courses. Rasmus had put a PDF document of his presentation on the website so that the students could print them out and have a preview of the lecture beforehand. Clara brought out the course handout and looked through it to have a brief idea of what Rasmus was going to talk about. Besides the handout, Rasmus had also attached three articles as literature, one of which was mandatory reading and the other two were supplementary reading. Clara read all of them on Sunday, because she felt that they were quite important and useful in providing a background and context of this area. This was helpful for her to understand the theories better by reading some case research.

The lecture[13] is in English, which, according to Clara, aimed at providing a chance for foreign students to participate as well as for Danish students to practice their English. All the group members were quite happy about this lecture, because firstly it was closely relevant to their project; secondly they planed to write the project report in English, which matches the literature and the whole lectures; thirdly, they could learn a lot from this lecture thanks to good teaching skills of Rasmus.

Rasmus started the lecture with an overview of the agenda for Module 5.

- Summary from MM4
  - Overview for position/speed sensors (special incremental encoder)
  - Principles for speed calculations with encoder
  - Overview of position sensorless techniques
  - Start-up procedures
- Comparison between motor/drives
  - Interest to fulfil
  - Parameters determining the cost price

[13]This lecture was about control of brushless machines, which is a PE course and will provide 1 ECTS. The course aim was to provide a comprehension of different control methods of brushless machines, such as permanent magnet motor, brushless DC machines and switched reluctance machine. It included 5 mini modules, which involved control of permanent magnet synchronous and brushless machine; control of switched reluctance machines; sensors and sensor techniques in drives; start-up procedures without sensors; comparison of drives. This is the last one of the five modules.

Normally the academic semester in autumn starts from September and ends at the end of sometime January after the exam. There are 4 periods included which last around 6 weeks each. In the first 2 periods, there are 6–7 times lectures each week which make up 50–60% of their work hours. From the 3rd period, most lectures would be finished so that students can use all the time working on their project. The project report is supposed to be delivered at the beginning of the 4th period, and afterwards students can prepare for different examinations for the lectures as well as the oral defence for the project report.

  - — Technologies there can be investigated
  - — Motor/drive comparison
- Case study: PMSM sensorless control (specific paper)
- Discussion and summary of course
- Exercise: step-motor competition

After the introduction to the agenda, he started a review of MM4. He showed the slide he used last time about the overview of position/speed sensors. In this slide, he used two diagrams — one is about the layout of slotted code disk, and the other one showed the direction determination with formula. This review was followed by the speed calculation with some examples. In illustration of comparison between motor and drives, he used different diagrams showing different elements regarding interest to fulfil, parameters determining the cost price, technologies that can be investigated, and a slide with different fruits and vegetables. For each slide, he showed the diagram first without texts, and then asked the students to think about the possible elements. After discussing these factors, he said, 'I know that it is extremely difficult to compare these, now think about how you compare these fruits/vegetables'. The next slide showed different vegetables and fruits. He gave students 2–3 minutes to think about it and discuss with the neighbours. Afterwards he showed 6 examples of motor types in comparison before he went to different tables for detailed methods for comparisons.

9:00, there was a five-minute break. Clara told me that the contents of this course can be quite boring and tuff, but Rasmus made it rather interesting. He was well-structured, logical and informative. Most of the slides he provided were clear, simple and visualized, which made it easier to understand. Clara said that she could not learn deeply by only listening to people talking all the time with text-based slides. Rasmus was quite provocative and energetic. He was quite aware of the concentration limits of students, so every other 10 minutes, he would provide some probing and facilitating questions. This made the course very challenging and exciting. Also he was good at bringing about different examples to relate the theories to daily experience, according to Clara, this way helps obtain a better understanding of the abstract theories as well as provide different perspectives of looking at things.

9:05, Rasmus continues the lecture with a case study of PMSM sensorless control. In this part, he used different diagrams and formulas to illustrate the theories. In this part, all the slides were filled with complex formulas with the process of calculation. He was speaking pretty fast and his voice was changing with rising and falling tones. Suddenly he increased his voice with smiling, 'hi, guys, are you scared at this complicated calculation? You should not get scared, you should enjoy!' Hearing this, students started to laugh.

9:45 Rasmus finished the lecture and had a summary of the whole course. He had a short evaluation by reviewing the expected goal of this course, and explaining how he tried to make it into a useful information resource and asking for feedbacks from students. A male student said, 'it has been the most interesting course this semester'. Rasmus smiled and replied, 'thanks, guys, I know that this has been difficult, but I hope that you have enjoyed.'

10:00, 15 minutes' break time. Clara told me that he was very precise on time, which was another point she liked about him. Clara and her group members went back to the room. I asked her how she felt about the lecture. She said 'the next half was quite complicated, it would be much more difficult if it was not Rasmus, I mean, he was very stimulating and impressive by his personality, which attracted people to pay attention, concentrate, and think deeply and understand more. He keeps our brain working all the time.'

In the project room, Christian was making some coffee for the rest. Carl brought out some bread from the shelf and cheese from the fridge and at the same time asked who else would like to eat something, Christian took one and Clara asked for a banana from the shelf. Jens was sitting in front of his laptop and looking at the screen. Clara was eating the bread while looking through the exercises they did in the last 4 modules. Rasmus was going to have a review of them as an end of the whole course.

The next half of the morning time was normally for excise. Lectures gave some exercises to students, who would be expected to practice in their project room. Clara said that it was the part she really learned things. She could also learn from the lecture time, but those things would drift away easily if she did not have a chance to practice immediately afterwards. By doing exercises that followed the lecture, she could have a chance to reflect on the theories taught and understand them better. The understanding could be promoted through discussing with group members and asking questions to the lecturer when he visited each project room.

Clara showed me the excise from mini module 4. There were 3 parts. Exercise 1 was a requirement to build or draw a 4-bit absolute encoder with a simple binary read-out (draw the encoder from 0–360). Exercise 2 consisted of 2 questions for drawing 'with what speed will it be good to change from method 1 to method 2 with the specifications in the course-slides? Use Matlab to draw curves.' 'What is the largest error with a combination of both method 1 and 2?' Exercise 3. A 6/4 SRM work without saturation with following conditions: n = 100 rpm, Udc = 80 V, D = 8%,. Phase resistance is R20 = 0.2 U. The current is assumed ideal square formed from 'corner position (15) 'to aligned position (45) with an amplitude on 10A. Calculate the position when the flux-linkage is 0.09 WB, with both a 20 c phase resistance and a 120 c phase resistance. Is there a big difference?

Clara told me that these exercise helped her to understand things better by applying the methods from theory to practice, especially by doing the third exercise. All of them involve drawing and calculation skills, which had been her strengths.

10:15 Time for doing assignments. They went back to the lecture room. Today's exercise was Step-motor competition + Finish all exercises from previous modules. After a short review of all the exercises from the first 4 modules, Rasmus announced the start of the competition. It was an assignment for each group to design a model climbing hilly surface for the speed calculation.

The competition was organized in the form of project groups. One or two students from each group would be the representative and the demonstrator. At the moment, male students turned excited. They were joking and talking loudly, trying to decide who should be the demonstrator. Clara was watching them and smiling. She said to me, 'this is really a guys' thing. They love to play with those technology things and they love to compete.' She told her group member that they could do it if they liked. I asked her why she did not want to do it. She smiled again, 'that's really men's thing. They mind so much those things. I don't care about that. I don't really like the technical things. So if they like, they will have the chance to do it.' I asked whether she could do it. She said, 'yes, I tried all of them and proved that I am able to do it, that's enough. I don't have special ambition in doing more with those.'

The atmosphere in the classroom turned quite high and exciting at that moment. Every group has three chances to demonstrate their design. The demonstrators were saying loudly that 'oh, I will make it, I will win this time.' They stretched the arms as a winner when they made it. When they failed, they

Picture 7.2 Clara's project room.

jumped and patted their own head and body to show the pity. The audience shouted to cheer up their group members, and whistled when other group failed. Clara and I was standing behind the group and watching. One of the other groups won eventually, and they were cheering loudly for themselves. Rasmus took three cans of beer from his bag and threw to them one by one through the air, saying 'hi guys, congratulations, here are the trophy for you winners.' Then he summarized the courses and said good luck and goodbye to the students.

12–12:30, time for lunch break. As normal, they had lunch together in their project room. They brought out food: brown bread, butter, cheese to the middle of the desks. Clara cut some slices of cucumber and put them beside the cheese. Carl turned on coffee machine and brought a couple of bottles soda water onto the table. They were sitting in front of their own desks eating lunch. Jens was eating his lunch when looking at the screen of his laptop. Christian and Carl were still talking about the competition, trying to analysing what was the reason they failed and what could have been done to change it. Sometimes lunch break could be a chatting time, but they were in a little bit hurry today, because there would be an extra lecture starting from 12:30.

They were supposed to visit a power plant in a week. This lecturer, Niels, offered an extra lecture for the aim of providing some background knowledge of this power plant. Clara and her group members told me that 'he was really a kind person, willing to give us his knowledge, but nobody can survive his methods.' They started to laugh together.

12:30, they went to another lecture room. Niels had been there, ready with the overhead on projector. According to Clara and her group members, he was one of the few old-fashioned lecturers who did not believe that students could learn from solving problems and doing project on their own. He insisted on his belief that efficient learning took place when students were given sufficient knowledge.

He provided different overhead with plenty of text based information. He looked serious when he was speaking. He kept the same tone and degree of voice in his non-stop speech. Half an hour passed, Clara said to me at a low voice, 'he is really trying to tell us all he knows, but look around the room.' I looked around and found that half of the 19 students were dozing and a few of them were reading some other stuff. Another half an hour passed, Clara told me that she started to feel tired now. Then she started to laugh because Niels asked a question to check how much they had understood, and there was no response. So he made a joke on himself, saying that he needed a 5 minutes' break. Clara said that he promised to give an extra lecture for an hour, but it seemed that it was going to be extended.

14:40, the extra lecture finished eventually. Clara's group walked back to the project room to get prepared for a short break before the meeting with their supervisor at 15:00. I asked them how they felt about the extra lecture. They started to laugh. Carl said to me, 'if you felt asleep, don't worry, because you were not the only one. We are used to him. He is actually quite kind, except for his belief on learning and his methods. But he is always helpful when he is needed. But you know, we want to do things in own way.' Just then, Niels came in and asked how they thought about the lecture, and they could ask more questions. Carl gave a cup of coffee to Niels and said that they were looking forward to visiting the power plant in reality to know more. Niels were about to discuss more things with Carl. Clara said to all of them with smile that 'shall we prepare a little bit before the meeting with David?' Christian and Carl said to Niels with joking tone that 'our manager said that we should work now!' Niels laughed as well and left. Clara told me that 'we need a person who can say oh, we need stop this and do the next one. You know, otherwise, these guys won't stop and we won't get things done.'

15:00, the supervisor, David, came on time. Clara made 6 copies of the meeting agenda and passed them to each of us. Christian gave a cup of coffee to him, and he took another chair in the room and got seated. Clara started the meeting by introducing the progress of the project since last meeting. Meanwhile, David asked some questions to specify some details. Afterwards, David gave some comments on what they reported, both positive and critical ones. Then they put some questions they had collected these days. Instead of providing answers directly, David responded by asking some questions as well. He asked them to put forward some options and write some formula on the blackboard, and then ask them how they thought about the situation. Then he commented on their replies. At the end of the meeting, they discussed what could be done in the following period of time based on the current situation.

Like all her group members, Clara liked this way of supervision, since she preferred to learn from thinking and getting to the point herself based on instruction rather than be given the facts and answers directly. David was very supportive and encouraging. Having his office close to their project room, he sometimes just passed by the group to see how it was going. They could also pass by for some small questions or spontaneous ideas. This friendship style supervision made both sides quite relaxed. For the group members, sometimes informal talks with supervisors could be of great help.

16:00, David left. Clara suggested a break before they continue with the work. She stood up to make the coffee. All of them stood up to stretch the body and started to chat something non-academic. The topics Christian and Carl were talking about were mainly about computers, sport matches, jokes. Jens joined this time for some discussion about a computer program.

16:20, they came back to the desks. They had a discussion about the supervision meeting. Clara brought about her notes from the meeting (they took turns to do that) and picked out some points for discussion. They discussed the critiques David talked about and agreed that there were mostly constructive and reasonable. They agreed to make some changed based on those critiques. As for the provided suggestions, they spent quite a while discussing them because some of them were quite inspiring, bringing more new perspectives into their project — this was the strength of David; However, some might demand some knowledge of some new areas, which should be carefully considered and discussed based on their time schedule. Eventually, they agreed that they should try some of the suggestions and give some of them up.

Then they started to modify the plan for this week based on the discussion today. Christian stood up and wrote the list of tasks on the blackboard, which included: finishing some chapters of the report, making some modification in modelling, doing research on some information about the new area they are going to add in. Then they divided the tasks. Carl suggested that one point of the tasks could be consulted in another group who had experiences before. He knew some guys from the group pretty well, so he could go there to get some information. Clara and Christian would work on the modelling and researching information respectively. Jens would continue with his part of the whole project, as they agreed.

17:00, Carl went out to another group for information. Christian and Clara started to work on their own work. The room became quiet. Once a while, Christian would say something to himself when he was clicking the mouse. Clara did not respond, because they had agreed on how to work together without getting bothered by each other, and they had been used to working together based on their experiences.

Being an engineering student involves long working hours, which was something they had been used to in the past 4 years. Normally at the beginning of the semester, they worked from 8:00–17:00 each day. This schedule would be getting prolonged little by little when they got into the later part of the project. In the past semesters, when they approached to the end of the project due time, working hours would extend to 20:00, 21:00 and even 22:00 in the last days. This semester they were working on a year-long project, so there won't be the semester report and exam, so they planed to have more regular and balanced work hours. They calculated the work amount and the new plan, then they decided to extend the work time to 3 days to 18:30, 2 days to 20:00, and from 10:00–17:00 on Saturday.

18:30, they agreed to finish the day and go home. Carl would do some shopping for their food storage this week. Clara reminded him to buy a specific kind of bread due to her allergy. They brought some work to do at home in the evening. Clare printed the handout and literature to prepare for the courses the next day, which was her plan for the evening. She told me that luckily, her boyfriend liked to do the cooking most of the time, which saved her some time doing other housework. However, her boyfriend was the real engineer type, who was very devoted to work and spent lots of time and energy with computers and electronics stuff. 'Very often both of us have to work long hours, and the children just have to learn to be independent and take care of themselves.'

When we were about to say good-bye, Clara and her group members had been working for 10.5 hours. They were going to work more individually later at home. I asked Clare how she felt about the day. She said, 'this is just one of the average days, there were so many things, but it is the life I have to accept as an engineering student. I am used to this.'

To conclude, this is one of the average days in the five-year-long engineering study life. It showed many common things as what had been discussed in the previous texts regarding the learning processes when studying EE engineering in the daily practice of PBL environment. Having worked in the same group for two semesters, the four of them got along very well. They could talk about many things, they laughed together, they understood each other's way of thinking, and they developed their group's way of working. For them it is first of all important to work with people whom they can get along with in order to benefit learning. However they also realized that by keeping to the same group they would miss new perspectives and critiques. Clara enjoyed this way of working very much, since she learned more through discussion and getting more perspectives to understand things and work out new ideas.

As for the aspect of studying hard-core engineering in the male-centred environment as one of the few females, Clara's experiences did not show dramatic differences from other female students and the female teaching staff whom I interviewed. It took Clara two years before she could have the courage to speak out her opinions in a group meeting. In the first year, very often she was not confident enough to speak out her mind, but rather sitting there listening to the male group members talking. Because she felt that she missed lots of technical knowledge that the male students were familiar with. From the second year, with the support from her boyfriend, she decided to change her situation.

Clara's story also illustrates a successful example. The improvement of professional skills brought about more courage and confidence. She managed to make her voice heard and make the male peer students believe that she was good enough. Clara became more and more brave and confident to speak out her opinions and she knew that her strength — being structured and good at management — had been realized and highly valued. At the beginning of June, Clara and her group members delivered their final report for the last project. At the end of June, they defended their thesis and were awarded Master degrees in Msc.Eg., which finalized their engineering study. In the celebration party, their supervisor suggested that they should develop the Master thesis into a journal article for publication, because it was an excellent project. They agreed to go ahead with it with the collaboration with the supervisor.

By then, all of them had signed contracts with different engineering companies. Clara was satisfied with the new job and was looking forward to the coming working life. Eventually she realized her dreams of (1) making a good living for her and her children by getting a well-paid job; (2) working as an engineer as her father did before he passed away; (3) working as the management level in an engineering company (she was promised this due to her pass experiences of working as project managers in her part-time jobs); (3) working as a female engineering (possibly manager in the coming future) so that she could bring women's values to the engineering work and management work.

In summary, the interview and observation data confirmed the assumption that doing projects in group work, attending lectures and getting supervision function as major resources of learning in the formal form, where both the cognition and affection of learners have been taken into consideration. In addition, empirical data also indicate that more resources for learning when studying in the PBL environment, for example, different social activities, which play an important role in sharing information as well as participating the generation of new information and knowledge. This learning process is a changing process for the learners towards active and self-directed learning, in which they learn to manage different facilitated resources as well as create new resources to make learning take place.

As most of the students informants reflected, the experiences of receiving education in this way was also a process of self-development whereby they grew up and became more mature. When students were at the first years, the aim of doing projects in group work was to be able to work together with others and pass the exam. Along the way when they did more projects and gained more experiences, the aim transformed gradually into a more active learning process, that is, how to develop strategies to manage one's own learning. Meanwhile, their professional confidence increased along the way.

7.5 Professional Identity Development in EE

Engineering represents a community of practice and has historically developed ways of practicing engineering. The structure of engineering learning culture encompasses academic training, technical language, social interaction routines and established practices for doing engineering work. The learning process involves a shared sense of cultural knowledge that gives meanings to the learners regarding who they are and what they do within the community. This is also a way to develop an engineering identity.

7.5.1 Curriculum and Devotion to Work

According to the informants, at AAU, the curriculum of different faculties[14] are normally identified by the schedule, work load and gender distribution. A comparison between the semester study schedule between an engineering program and a social science programme (can be seen in Appendix 5) show that there are more fixed arrangement like courses and activities in the engineering curriculum.

The daily schedule in most of the engineering programs starts from 8:15, whereas in social science and humanity faculty, it often starts from 8:30 or 9:00. Among the students, there are some ways to distinguish students from different faculties. In the morning, buses arriving at campus around 8 o'clock are normally full of engineering students who wear mostly classical wind coat, jeans and the practical shoes, and mostly guys. As one student said, 'If one day you are late and come in the bus one hour late at around 9 (even more), you would find yourself different from the rest people in the bus. They are mostly from social science or humanity areas, where courses start after 9 o'clock. They would be more fashionably dressed, talking loudly and there are much more girls around.' This none scientific joke might sound exaggerating, and when talking about this, students were laughing, however, it showed a way people identify themselves culturally based on their disciplines.

With respect to work load,[15] most of the students said that they worked from 8–10 hours at the beginning of the project and extended to 15 hours and

[14] At AAU, there are three faculties: Engineering and Science Faculty, Social Science Faculty, and Humanity Faculty.

[15] With respect to work load, this study did not make efforts in making comparison between students from different disciplines.

even more when it reached the end of the project. During the interview and observation, I did not hear any complaints about the work load. According to most of the interviewed students, studying this way was regarded as an imitated engineering work life. It was what they knew about the real engineers' work, therefore, they would like to train themselves to get used to it so as to prepare themselves for the workplace.

When asked about their spare time activities, male students talked about their obsession to computer games during adolescence time and the first a couple of years at the university. However, most of them stopped it when they felt that 'they became more mature' and would like to spend more spare time doing something else like sports and music. Female students mostly referred to activities that were different from engineering, swimming, badminton, sewing, painting, yoga. However, all of them said that it was very time consuming to take engineering education, which occupies most of their time, especially in the first years when they had to do extra work to catch it up, they hardly had spare time for these activities. During every project, there would be two months (around project due time and exam time) almost without social activities outside of university life.

Working on projects sometimes means that there were no regular work hours sometimes. The 8–16:00 schedules only worked at the beginning of the project. When it reaches the project due time, the more working hours it might involve every day. In the last 6–8 weeks they would work until 9 or 10 o'clock in the evenings. In one group discussion, the male students mentioned that public holidays were not for engineering students. They hardly had holidays throughout these years,

> "... no time to relax in Eastern days in spring, no autumn holiday, sometime exams at the beginning of January meant lots of work between Christmas and New Year Day. Summer holidays might be the only time to relax ... Sometimes we can't notice holiday until we realize that we have to use the special entrance key ... You know in this building, we use the card as key to the entrance, while in holidays we have to use the code. Then sometimes we got to know 'oh, it is holiday

Picture 7.3 Engineering work at EE.

> today,' only when we were asked to dial the code. That's how we got to know holiday."
>
> (Male group)

As they reflected, the main reasons for this hard work are due to the personal ambition and interest. Working on projects will unconsciously increase work load, when they get more and more involved, they tended to spend more time and energy on it. And none of the interviewed male students regarded this as a problem. According to them, this was not more difficult than their expectation. As Carl said, 'sometimes it is not that bad when I got deeply involved in the work. I don't really realize that 'oh, I am using all the weekend working in the university', so I think that it is ok.'

This study did not make efforts to compare the work load of students among different faculties. Therefore, it is difficult to judge whether it is harder or not to get through engineering education than other education. However, from the perspective of EE engineering students, it demands strong principles, sober attitude and lots of efforts to get through the study in an engineering program, because it will lead to a serious profession. According to them, this way of identifying engineering study has been agreed by students from different faculties. Also agreed by these students were that studying in the PBL environment, it helped get through this education since working in the groups; they were not alone to face difficulties.

7.5.2 Engineering Ways of Thinking and the 'Nerd' Image

When describing the general engineering ways of thinking, students of both genders shared similar vocabularies: 'problem-solving oriented', 'focused', 'structured', 'logical', 'rational', and 'analytical'. According to them, these words are closer to the ideology of male in the society. This is also one way engineering students distinguish themselves from students in other disciplines.

> "At the university, people who are studying non-engineering are not as structured as we are. You will be that way (well-structured) when you are having technical education."
>
> (Juli)

'Engineers are nerds!' This was a rather common way for engineering students in EE to tease themselves through all the interviews and observation. 'We are engineers, we are nerds.' These words were often heard from male students during the research. They were clear about this image before choosing this education. All the interviewed male students mentioned that there was a period of life when they were 'really nerdy by spending most of the time in front of the computer'. This experience was regarded as 'one of the necessary period on the way to becoming an engineer'.

On the other hand, none of them agreed that they were 'socially incompetent'. According to them, lack of social skills is a general comment on engineers by non-engineers. However, working in groups involves plenty of chances for communication, as some of them said that this was a good strategy for engineering students to avoid isolating themselves from social activities by sitting alone in front of commuters'. By working in groups, they unavoidably participated in different social activities.

Female students were also aware of the nerd image of engineering students before they entered the profession. However, none of them associate this image

Picture 7.4 Project room of a male group at EE.

to themselves. According them, the word 'nerd' is more associated with male image and distant from femininity.

7.5.3 Interactions and Communications

Interaction and communication styles play a role in the construction of the learning culture. Gendered patterns are especially visible in the settings of daily practice in EE.

Male norm in interaction

It has been natural for male students to enter engineering profession, the image of the profession match their gender role, the contents they study fit into their interest and background, therefore, they feel at home in this male-centred environment.

In EE, with men's participation as over 95%, a strong masculine culture is indicated through physical surroundings, discourses and communication patterns among male teaching staff and students. A presentation of masculinity can bc observed through daily practices: high visibility of men around the building; in the project rooms: porn pictures on the wall and beer bottles on the floor; technical terms in jokes, aggressive ways of conducting discussion and arguments, etc.

Interaction between students and the teaching staff

There was little interaction between male students and female staff due to the general underrepresentation of women employees as teaching staff in EE. Interaction between male students and male teaching staff developed from pure professional contact in the first years to both professional and informal contact as friends when they reached the last years. It had been a natural process for male students to get along with male staff when working in the male-centred environment.

To different degrees, the interviewed women students in EE experienced gender bias in the first year. They sensed that some of their male peer students and professors did not welcome them, or did not take them seriously as would-be engineers. As new comers, they bore heavier burdens than their male peers did. Given the lack of technical knowledge in the background at the starting point and the added pressure in the university, these female engineering students had to deal with this pressure and insecurity as they struggled to get through the demanding engineering program.

In the later years' education, most female students experienced 'equal' treatment with their male peers by teaching staff. They did not want to be treated differently, because being different could be interpreted as being not qualified. As Line said that she would like to keep professional and strict relationships with supervisors, 'because they would be the examiners. I wanted to be judged fairly by my capabilities rather than relationships.'

These female engineering students referred to their preference to be judged the same way as their male peers because they would like to prove that they were professionally qualified. Mia mentioned some rumour that male teachers would give privilege to female students in exams,

> "Some people said that as long as you are a girl, professors will let you pass. There are two reasons, one is because they show mercy to girls, and the other one is they don't take girls

> seriously. I don't want to be judged that way, so I prefer to be treated the same as guys."
>
> (Mia)

Based on the strong wish that they wanted to be accepted, these female students worked hard and tried to prove to the professors that they could be as good as their male peer professionally.

Most of the interviewed male teaching staff had little experiences of teaching female students, and they chose to treat all the students in the same way. However, Freda's experiences showed that some privileges from professors can be positive and supportive.

> "I met some very good teachers; they gave me quite some privileges in the first years. They were so warm to give me help in technical part; that really supported me a lot to get through the hard period. It prevented me from dropping out."
>
> (Freda)

The female engineering students in this research showed their strong mind to prove their capability in order to be accepted. The general academic skills of these female students, especially their proficiency in math, were highly valued by their supervisors and male peers. However, special support and help from teaching staff in the early period could have helped shorten the time and reduce the energy that has been taken to pass through difficulties.

Interaction among peer students

Male domination brought about a feeling of being different to the female students at the beginning, which led to a sense of insecurity, especially when they are minorities. In order to be taken serious at work by their male peers, they take special consideration of their dress and manner.

Communication with male peers was difficult for female students in the first years. Most of them had a hard time being accepted from the beginning. They were suspected by the male teaching staff and their male peers with regard to whether they entered the right place. Seven out of the eight women talked about their experiences of 'voice could not be heard' in the first years. It was also the young and naïve age for male students, who directly expressed their bias towards women students who showed interest in technology but might not be as competent as them in terms of technical background. Both Freda and Laura experienced some guys in their project groups who showed their bias towards women directly. When they expressed their opinions in the group work, the guys were using the same biased tone saying to them that 'you girls don't know this area' or 'how can you girls do it this way', etc. Facing this, Laura felt very sad and inconfident, 'he made me feel I was so stupid because he knew something that I did not know.'

Freda developed some strategies when she moved to the third year. She tried to argue with the guy and proved that she was good enough,

> "I got so bothered by his attitude, so I argued against him. I said 'how can you say this, we have the same education here, we are learning the same things, I can do what you can.' I also proved it with my action. By doing this three and four times, I managed to make him shut up."
>
> (Freda)

After this, Freda learned to develop some strategies to avoid problems like this. Since then, at the beginning of every project, she would frankly talk about her attitude and expectation to the male group members.

> "I told them my strengths and weakness as well as things I would like to learn through this project. Also I told them that I would like to be respected because I do contribute to the project work. By telling them that bad experience I had before, I hope that it would not happen again. Then it worked quite well in the later project."
>
> (Freda)

Having similar experience, Sara chose another strategy: she invited another girl to work in the same group. By doing that, she felt more confident when talking in the group.

> "I felt much better when there were two girls in the same group. I know that there will be at least one person listening to me when I talk. It helped me to gain confidence."
>
> (Sara)

Being the only female in the group and working with 5 or 6 male students in the project, female students are facing the male language and jokes. However, they felt strange if they asked the guys to stop their male joke 'because of the presence of a girl'. Therefore, they chose to adapt to the majority rather than to ask for change.

> "I will feel that I am more strange and different if the guys have to stop their jokes (dirty words) because of me. I want them to do it in their way. They are guys, they are that way all the time."
>
> (Maja)

Mia's strategy made it easier for her to be accepted, and she did not mind being 'one of the guys',

> "I had no problem handling those guys, my way is just to talk about their topics, and to have more dirty mouth than them, then we get along very well ... I know that they don't regard me as a girl, it doesn't matter ..."
>
> (Mia)

All of them mentioned that they had got used to it so that they did not notice them any more. It took a couple of years for these women to move from 'feeling being different' to 'feeling being special'. This change is an evidence of increase of confidence, because they started to see their own value and privilege that can be achieved. As some of them said,

> "I think that it is quite fine to be special. There are some privileges: everybody around will know me, which makes me feel good sometimes. Then it is easier to get things done when everybody around knows me."
>
> (Juli)

Working in the male-centred environment for years, these female engineering students developed different strategies to be accepted and to provide more chances for themselves to learn in groups. When they went to later semester of the education, they became more experienced in handling different situations as the minority; they also learned to cooperate well with male peer students and developed good friendships with their male peers. Some of them won the admiration from their male group members, who could appreciate their values and contributions. As both the supervisor and male group members of Clara said that, she was such an intelligent woman that they all need turn to her for technical help.

> "Also, we enjoyed working together with her in the same group for years, because of the personality she has and the good atmosphere when we work with her. It makes us enjoy learning together so much."
>
> (Christian and Karl)

Both Freda and Clara talked about the supports (both mental and technical) they got from their boyfriends which helped them to get through the most difficult moments and kept them from dropping out. In addition to the bad experience she had from the immature and hostile male group member, Laura also mentioned some helpful group members she worked with in another project,

> "This semester I have some very good group members. Some of them helped me a lot, sometimes they explained the things to me with patience. I felt that I learn a lot from the group members than even from lecturers and supervisors."
>
> (Laura)

This evidences that male students can be of great help to the female peer students, if these men have the awareness of gender difference as well as the intention of welcoming women to engineering.

Having the feeling of being isolated, and having few role models in the department, these female students expressed their wish to have more female peers. There were generally two reasons for this wish: (1) they needed mental support from each other which will help them gain more confidence to speak out their opinions in professional activities (for example discussions in group work); (2) they wished to have a different culture from the male-centred culture so that they could be 'social in a girl's way'. They can talk about things they are interested in instead of listening to the guys' topics all the time (like sports and computers). As Clara said,

> "I liked the feeling of being a woman when I once in a while hang out with a female friend, it reminded me the feminine part of myself..."
>
> (Clara)

In summary, studying EE engineering in the PBL environment, students developed a sense of engineering responsibility and engineering work styles. This research also witnesses a gender pattern in this professional identity development. Some features in the learning processes, such as solving problems and doing projects in team work are manifestly recognized as strategies of building up professional confidence and belief for career pursuit. However, this professional identity development is less evidently recognized by women students, for whom, it was the education they were going through based on their interest in science.

7.6 Students' Perceptions on Gender Relations

As has been reported in the last sections in this chapter, gendered patterns in the learning processes in EE are rather visible. During the course of interviews, female students were more aware of the gender issues[16] and sometimes brought about the topic by themselves when they talked about their experiences in the learning processes. This seldom happened during the interviews with male students. In most of the interview cases, I brought about the questions on gender after the discussions of their experiences of learning. In the following, I will report how the students in EE engineering in my research perceive gender roles, masculinity and femininity, based on their life experiences.

7.6.1 Gender Roles

The discussions about gender roles in the current Danish society reflect a visible change from the parental generation of these students to their own generation. More than half of the interviewed students mentioned that their fathers had jobs in traditional male professions, such as working as engineers, technicians, farmers, etc., whereas their mothers had jobs in traditional female professions, such as working as nurses, secretaries, accountants, etc. Most

[16] This might also because they knew that I had a gender perspective in my research beforehand.

of them grew up in families where there were traditional gendered labour divisions: mothers were responsible for traditional housework for women, like cooking and cleaning; whereas fathers were responsible for traditional male housework like repairing houses, doing electrical work, etc. For those male students who grew up in this family background, it is natural for them to help their fathers with the electrical, mechanical and repairing work since childhood, which is something the female students seldom experienced.

Many social changes have taken place in the past decades in Danish society. Many students talked about the equal situation in the gendered labour division in the family life in their generation. Most of them who had girlfriends or boyfriends share the housework. In spite of this, most of the male students said that they preferred to have girlfriends who were not from engineering area, and who were not necessarily well-educated.

7.6.2 Gender Identities

Students of both genders in EE share similar ways of perceiving the influence of gender. Outside appearances, ways of dressing, using languages and manners were regarded as important aspects to distinguish men and women. Professions remained another important factor that divides men and women in the society. For example, some typical female occupations are: nursery, teaching in primary school, hair dresser, waitress, shop assistant, cashier in the supermarket, and so on. Some typical male professions are soldiers, construction workers, technicians, engineers, etc.

When talking about masculinity and femininity in the current Danish society, both male and female students in EE brought about some vocabularies. For masculinity, most of the students of both genders referred to the terms of 'logical', 'decisive', 'analytical', 'responsible'. When speaking of their opinions of femininity, the first words most male students brought about were 'gentle', 'sweet', 'elegant-mannered', 'intelligent', 'independent'.

For the male students, it was natural for them to become engineers since they were doing the suitable thing for their gender role in the society. For the female students, it brought about a dilemma. They had the awareness that engineering profession was not suitable for the female gender role. According to them, choosing this non-traditional occupation involved hiding some parts of the femininity. For example, they mentioned the consideration of ways of dressing at work. Working with engineering guys, they need be very careful about their ways of dressing. Feminine clothes like short, tight T-shirt and skirt might bring biased eyes from their male peer students. As Laura said,

> "I don't want to be dressed too feminine, because they (the guys) would think that I want to use my body to hide the lack of intelligence. I want to show them that I am professionally qualified..."
>
> (Laura)

Working in the male-centred environment for long, these women felt that they tend to think (to be more logical, analytical and structured) and talk (to be more straight-forward and simple) in men's way. This made them different from what they used to be and also different from their former female friends, who mostly worked as hairdresser, primary school teacher, nurses, animal doctors, or students majoring languages, psychology, and other subjects in humanity or social science.

Sara used the words 'here' and 'there' to distinguish female students in different faculties. Engineering girls were from 'here'. Female students who study humanity or social science were regarded as 'from there'. For Sara, 'girls there' were quite different from 'girls here',

> "Sometimes I feel that girls there come from another world, the way they talked and the things they talked about are so different from girls here. Maybe it was because they are doing soft study, and we are working on hard stuff. Anyway, I just feel kind of difficult communicating with people who are not from the same world..."
>
> (Sara)

Most of the female students in EE enjoyed being different. However, some of them would like to keep femininity at the same time. Both Clara and Freda talked about 'to be a feminine engineer'. In Clara's opinion, when working as engineers, women might change into more men's way and get to be more different from other women; however, it should be possible to be soft and sweet in manner and at the same time of being an engineer. 'You don't always have to be tough, cool and controlling when you work as an engineer. It is not necessary to behave in a man's way to prove that women can be higher than men. Equality is good enough for me.'

Summary

In this chapter I have reported the empirical study of my research in EE. This effort was an intention of portraying the learning culture of EE in PBL environments based on the experiences of students of both genders. In general, the PBL environment plays a positive role in the learning processes of students of both genders. They learned to take responsibilities of their own learning, to take active participation as well as initiate different activities for the purpose of creating learning opportunities to achieve their learning goals.

This chapter also discusses in detail how students experiences the learning processes and the development of engineering identity in gendered forms through everyday practices, interactional activities in EE. The gendered patterns in entering engineering field, taking unequal starting points due to their different technical background in their past experiences, men and women experienced different learning processes in studying engineering. The nature of hard-core engineering subjects that is based on male interests and their past experiences brings about privileges to men and serves as barriers for women in the learning process.

8

Studying Architecture and Design Engineering

This chapter reports the empirical work in the Department of Architecture and Design (A&D) Engineering. The depiction starts with a brief introduction to the study programs in A&D, student's choice of engineering and the specific study program, and their experiences of being newcomers. Followed by is the report of their perceptions of learning, and the learning processes of studying engineering in the PBL environment through doing project in groups, attending lectures, gaining help from supervision and participating in different social activities. This learning process also witnesses a development of professional identity through participating in the professional activities as well as social activities. Also described is the learning culture in the program of Industrial Design in A&D from a gender perspective regarding gendered proportion, gendered features in the learning processes, and students' perceptions of gender roles and gender relations.

8.1 A Brief Introduction to the Study Programs in Architecture & Design

This section provides a brief introduction to the Department of Architecture and Design with focus on its origin, learning objectives, and curriculum design. The data resources come from the following sources: relevant documents and websites[1]; interviews with the head of the study board, who participated in

[1] See http://www.aod.aau.dk/

Gender and Diversity in a Problem and Project Based Learning Environment, 201–264.

the initiation and organization of the program, and who had a background of architecture; interview with head of the department, who is responsible for the overall management of the institute, who has a background of organizational management; interview with two semester coordinators, who are respectively responsible for the 6th and 7th semester curriculum arrangement in Industrial Design, and who both have a background of industrial design; documents from the department.

The Department of Architecture & Design, Aalborg University, established in 1997, provide 5-year-long study programs which offer Master degree in science and engineering (M.Sc.Eng). The programs contains 6-semester-long Bachelor level study programs, which are held in Danish and 4-semester-long Master level study programs, which are held in English and open to international students.

The study programs in A&D intend to build up a new study area that emerges from the technically oriented civil engineering programs and the artistically oriented architecture and design programs. It provides study chances for people who want to create, design and innovate, and people who are interested in technical matters and information technology. The program is open for anybody who fit into the general admission requirement of the university. There are no compulsory demands for IT and drawing background.

The objective of the study programs in A&D is to train students to have

> "an advanced and creative combination of technical and aesthetic skills in problem-solving and modelling, and an ability to balance often opposing functional, production-related, economic, and environmental demands and interests, and a thorough knowledge of the financial and social conditions of the occupational function and an understanding of the social and health consequences of different solutions and designs."
>
> (A&D Study Regulations, 2004: p. 7[2])

Curricula in the study programs in A&D (see Appendix 2) include two semesters (the first academic year) of basic core courses; three semesters of advanced core courses and five semesters of specializations in Urban

[2] Information source is from http://www.aod.aau.dk/uddannelse/studienaevn/index.htm

Design (URB), Architecture (ARK) or Industrial Design (ID). During the sixth semester a Bachelor project is written within the fields of URB, ARK, ID. The Master program ends with a thesis project within the same three fields plus a new specialty Digital Design, which started in 2003.

The basic courses are provided in the 'freshmen building,'[3] which covers elementary skills and knowledge in studying, methods, and tools. Students at 3rd–6th semester are expected to acquire a wider knowledge and professional competence regarding design methodology, technology, aesthetics and communication, and carry out design projects on different scales. In the advanced courses, students are taught the main methods of architecture and design as well as core subjects related to aesthetics, form, function, construction and materials. Students will select their specialty after the fifth semester so as to get prepared for professional tasks.

PBL concepts have been implemented in the study form at A&D since its establishment. According to the head of the department and the head of the study board, the PBL concepts are practiced in diverse ways in different semesters and specializations at A&D. Every year, modification is made in the implementation of this educational model based on the experiences and responses from both teaching staff and students. In general, each semester students carry out a project on a specific topic. The project work comprises half of the total work time. The other half includes lectures, exercises and a field trip. Since the 4th or the 5th semester, project units in A&D fall into two categories: main projects which are normally carried out as group projects; minor projects which are normally carried out individually or in exceptional cases in terms comprising a maximum of two people.

Staff members who are employed for the teaching tasks in A&D have different professional backgrounds, for example, architecture, industrial designing, mechanical engineering, and so on. Most of the teaching staff who have from non-engineering background used to work in different other educational institutions outside of AAU. Few of them experienced PBL neither as students nor as educators. Interviews with the teaching staff in my research witnessed generally positive response towards the practice of PBL concepts at A&D.

[3]This is my translation of the name from Danish to English. In the Engineering and Science Faculty, AAU, all the first year students study in the shared buildings, which are in a separate campus from other parts of the university.

There have been around 100–150 students enrolled every year (with an increasing number of applicants each year), and there are currently around 500 students (with around 50–60% female) and 50 staff members (with around 30% female) in 2005. The first generation master level students graduated in 2004. By the end of June, 2004 before I finished my fieldwork in A&D, all the students specialized in Industrial Design at 10th semesters found jobs; most of them were employed in different companies in Denmark.

In this study, information from interviews with the teaching staff is mainly used as background knowledge rather than the focus of analysis. Therefore, this section aims to provide a brief introduction to the research site at A&D from the perspective of educators. In the following sections, I will move to report the research findings based on the experiences of students who were at 6th semester in the specialization of Industrial Design in spring, 2004.

8.2 Path to Engineering

This section reports the reasons for these students to choose the engineering study in the program of Architecture & Design at Aalborg University. Data are mainly drawn from group interviews and informal talks with 30 students (21 female and 9 male students) at the 6th semester during my research in the spring semester, 2004.

8.2.1 Choice of A&D

As a new educational institution that has an image of design, creativity, and technology, Architecture and Design (A&D) attracts more and more students each year. The interview data in this study identify a similar path to engineering among all the informants of both genders. When asked why they chose this study program, the responses were identical: because of the combination of design and technology.

The majority of students said that they were interested in doing something creative, for example, designing, and they would like to do it with technology, however, they did not want to only focus on technology. Thus, the study programs in A&D provided a good option for them.

> "I chose it because I was interested in design and technology, and I would like to do something creative with technology,

> and this program sounded interesting because it is a combination of two kinds of education."
>
> (Nathan)

> "I chose it because I like math. I like to do something with technology and engineering, but I also like to be creative and do things in new ways. So this education provides a chance to do both."
>
> (Mary)

Besides the above-mentioned reason of interest, there were three female students who chose this program because of their dream about the profession of design or architecture.

8.2.2 Choice of AAU

When asked about the reasons for choosing this university, the most cited response was that it was the only place that provides this kind of program covering the two different areas of design and technology in Denmark.[4] Some students mentioned two other options to receive high education based on their interest at that moment, Aarhus school of Architecture and design programs at DTU, Copenhagen, both of which provide study forms with focus on individual work. They did not choose them because the former had little technical concern in the teaching contents and the later was too focused on technology.

Among the total 30 informants,[5] 22 of them had a background of mathematical gymnasium. Few of them mentioned the study form as a reason for choosing this university. For the 7 students who graduated from technical gymnasium, their experiences of studying technology in the form of doing project in groups made it natural for them to continue higher education at an engineering university, especially AAU, where a similar study form was provided. Among them, the five male students referred to their interests in design and creativity as the main reason for choosing this program instead of other engineering programs. With their tinkering experiences in childhood, earlier access to computer, and passion of technology, they expected

[4]This was the situation when these students started university in 2001.

[5]For more information about informants see Appendix 3.

sufficient technical and engineering contents in the teaching materials. As they said,

> Coming from technical gymnasium, we had some basic knowledge about technical things, we expected that there would be more advanced technical knowledge here… besides, we wanted to have something else more than pure technology, for example, like design, and how to be creative…'
>
> (male students in group 2)

In contrast, the two female students chose this program because of their wish to study engineering with a reduced technical contents. As one of them said,

> "I would like to learn more about technology, but it is boring to only focus on the technical things. I chose this program because it covers technology, but it is closer to real life."
>
> (Louisa)

For these students in A&D, family played a supportive role, but not a guiding role with respect to their decision making. The majority of the informants have no family members who work as engineers. Their decision making was mostly based on individual interests. The three of them whose fathers are engineers said that this family background might have aroused their interest in technology, but did not directly influence their decision making.

8.2.3 Pre-knowledge About Engineering

The process of choice making for these students was also a process of exploration. They wished to receive higher education based on their personal interest, they got information about this new study program from friends or other acquaintances, they looked up introductory materials to the teaching contents, and they talked with study chancellors and family before they made the decision. They were attracted to make the final decision on this program for the strong interest in working with technology in a creative way rather than studying engineering itself.

Their impressions of engineers before they started the education varied. Nearly half of the students had few ideas about this profession — who engineers were and what engineers did. The other half of the students came out with words like 'technology-oriented', 'nerd' and 'not creative'. In general, few of them would associate what they were going to study with engineering.

8.3 Being Newcomers

When these students were enrolled to the Department of A&D in 2001, there were no special requirements for computer and drawing skills from the perspective of the institute, in that courses for these basic skills would be provided in the first year curriculum. In this way, as novice who stepped into a newly established organization, students in A&D had rather equal starting lines when they began to develop new competencies.

The interview data in the program of A & D show that the students' experiences in the first year involved a transition process; nevertheless, without clear evidence of gender differences. The following are the main issues that were discussed throughout the group interviews when the students described and reflected on their experiences as newcomers in the first year. These issues came out of interviews with all the groups, though specific concerns were stressed more in some groups than in others.

First, confusion and disappointment arose from having no clear information about the learning objectives. Being open to new things, these young people took challenges in life by choosing a newly established study program at a university without available concrete knowledge about the learning contents. They were attracted by the program description, which indicated to them what the curriculum was designed to be, however, they did not have clear clues about what it would be in reality. As described in the discussion of one of the group,

> S1: "When we started, we only knew that it was going to be a combination of two things, it sounded very interesting, but we did not know more than that… "
>
> S2: in my imagination, it was not only either technology or design, but a combination of them, like real life, but we did not know how things would be combined…

> S3: it was a new education at that moment, and no one graduated yet, so we could not know very much about it . . . (Group 4[6])

As newcomers, they started the university life with ambiguous imagination about the new life. They were looking for signals and indications of the learning objectives and expectations from the study program. Some students also mentioned their confusion from not being informed of the learning purpose of some courses and the expected learning outcome.

> S1: "(in the first year), we were offered almost the same courses at different programs in engineering faculty. It was difficult to see why our program is different from other . . . "
>
> S2: "We learned lots of things that we could not really use later. Sometimes the teachers were talking a lot, which only confused us."
>
> S3: "They might be the masters of those areas, but it did not help when we could not see how the subjects could be used and how they could be linked to our project. For example, like mathematics, which was quite hard to learn at that moment because we could not see how to use them."
>
> (Group 3)

Thus, for most of the students, they fell into confusion at the beginning of the education by having little understanding of the purposes of the teaching contents, having few ideas about the learning objectives, having little information about the expectation of their learning outcome from the department.

Second, the physical environment brought about difficulties for the students in A&D to get used to the new student life. In the first year, they were situated in the building for the entire first-year students in Engineering and Science Faculty, which was a different physical environment from the buildings of

[6]In this chapter, the origins of citation from interview records fell into two forms: (1) with group number in the case of quotations from different interviewees in group discussions; (2) with individual names (pseudonyms) in the case of quotation from specific individuals during group interviews.

Department of Architecture and Design. Every group had a separate room for project work, which reduced the chances for students to participate in social activities. As students in one group said,

> "We were not social in the first year, every door was closed. And nobody wanted to knocked at others' door and see what others were doing."
>
> (Group 5)

According to the students, this way of arranging study places made it hard for the first year students to get to know each other and to build up communicative atmosphere. In stead of providing students chances for sharing experiences with each other, it provoked negative competitions among different groups. The feeling of being isolated in the first year was expressed in most of the group discussions. As one student said,

> "I felt kind of isolated. We were locked in small rooms for group work, so it was hard to get to know more people. I really hated that way."
>
> (Joan)

The unsupportive physical environment plays a negative influence on the learning process of the students both psychologically and socially. This situation did not change until the second year when they moved to the building of A&D department, when the physical learning environment changed from separate project rooms to open studio.

Third, it was time and energy assuming to get accustomed to the new learning environment, new study form, new subjects and new methods of studying at the same time. The majority of the students in this program graduated from mathematical gymnasium, where ways of study were mostly in the form of reading, attending lectures and doing individual assignments. Having little experience in doing projects in groups, they spent lots of time getting used to the new way of learning. It also took time to be able to reflect on their experiences and sensed the meanings of this new way of learning. According to the students, it was especially hard to see and feel about the learning outcome at the beginning. As said by a student,

> "At the beginning, it was so difficult to see that you were learning anything just by sitting here talking, because in the gymnasium, we learned mainly from attending lectures, and suddenly we are just sitting here, talking and discussing with 5 or 6 other students. It was difficult to feel that I was learning and that was the right way to learn, but I think it just takes some time to get used to this way of learning."
>
> (Michael)

Working in big groups (with 7–8 students in one group) in the first year made it especially difficult for the novice to learn to manage the project work in groups. All the groups talked about the difficulty in reaching agreement on what to work on and how to get work done. Sometimes they made comparison with their friends who studied in other universities (with different study forms) and other programs in AAU (in some traditional disciplines). Their friends either study by reading lots of books or knowing clearly what they were studying, these students felt 'empty by just sitting there and spending lots of time arguing just in order to agree on some small things'. Thus, in the first months, they could not achieve the feeling of learning, which increased the insecurity.

Fourth, heavy workload led to stress when confronting different new things at the same time. Life was occupied with preparing for lectures, attending lectures, doing project work, making efforts to reach agreements in the group work (based on different ways of working and different personalities). As one student said,

> "We carried lots of stress in the first year. I remember that we very often had to work longer than people from other programs. When they worked maximum from 8–16:00 in the first year, we worked minimum 8–20:00. It was because our project was so broad so that we had to collect lots of knowledge to make the project. In other engineering programs, it was more focused ... "
>
> (Lee)

Similar stories were heard in most of the group interviews and informal talks. The heavy workload brought about imbalance in life, and played an unenthusiastic and unhelpful role in motivating novice to learn.

Fifth, this new way of study life brought about changes in life, which in turn, influenced the private life. Most of the groups talked about how changes took place in life since they started university, which brought about difficulties in keeping the balance of work and private life. Leaving home to start an independent life was exciting; however, when life was occupied by getting adjusted to new study life, they hardly had spare time visiting family and friends outside of university. As a female student said,

> "At the second semester I started to wonder whether it was the right thing for me (to continue with education). We had to work so much and I was afraid that I would lose my friends and family when I could not spare time seeing them. I needed spend time staying with my boyfriend, visiting family. I need energy from outside."
>
> (Lea)

Quite a few of them experienced breaking-up with boyfriend/girlfriend in the first year, because, according to them, it was difficult for the other part to understand why they had to work so much when studying at this program.

Having difficulty finding the meanings of learning, 35 out of 130 students[7] dropped out after the first year. The main reasons, as analysed by the rest of the students, might be the roughness in the study, as one student said,

> "I think that most of the students chose to stop because it is hard to define our education, what we are going to be, where we are going to work, what job it will be, all these things. And we were not sure what we had learned in perspective of technical knowledge… "
>
> (Anta)

Therefore, the experiences of students in A&D in the first year was a transition process, in which they learned new subjects, new ways of learning by doing projects in groups, and new ways of living through becoming university students. This transition is also a difficult process, which involves confusion

[7] In this study, the information about dropout came from the interviewed students in this study. The officially registered reasons for dropout were not accessible during my research.

coming from unclear learning objectives led to uncertainty in the learning processes.

8.4 Learning Culture at A&D

This section reports the learning processes of these students in A&D who chose the specialization of Industrial Design at the 6th semester. Data are drawn on both group interviews and observation in my fieldwork. During the interviews, students talked about their perceptions on learning, different learning resources, and how they experienced the study life in the PBL environment in this newly-established study program. Also provided in this sector is a portrait of the process of carrying out a semester-long project work based on my field notes.

8.4.1 Students' Perceptions on Learning

For most of the students, it took them two or three semesters to get used to the university life. Once they got through the transition process, as some of them said, 'when we find out that this is what we want to do, we may get less stressed because we are on the right track.' Since the 3rd or 4th semester, when students looked back on their past experiences, they reflected on what they had learned.

> S1: "we discussed a lot about the issue of learning. In the first year, we had a hard time measuring how much we have learned when there were so many things . . . "
>
> S2: "it was kind of conflicting, on one hand, we knew that we were learning all the time, one the other hand, we realized that (what has been learned) mostly afterwards, after the project for example."
>
> S3: "sometimes when we are talking about what we can do in this project, we realized that oh that is what I have experienced in the past, and that was what we have learned, then we can use the experiences in something else later."
>
> (Group 1)

Once students started to realize that the past experiences (though involving lots of difficulties and confusion) could contribute to the achievement of learning in the later experiences, they started to see the strengths of learning in the PBL environment.

Physical study environment played an important role on building up a sense of belonging to the study program. Since the 3rd semester, they moved to the building of A&D, where an open studio was provided as their shared project room. Every group got a part of the studio which is separated from each other by the boards, shelves or cabinets. This open environment provides a friendly social atmosphere, which helped students share information and improve communication not only in the group but also among different groups. As discussed in one group:

> S1: "We started to see the good point of learning this way (in PBL) since the 3rd semester, when we moved out of those small rooms."
>
> S2: "this open study environment is really a good idea, because you can see how others are doing things, you can hear what other groups are talking about, you can ask questions."
>
> S3: "It is more social. We share information and we learn from each other — We know sometimes it can be very noisy, but it is better than working in small isolated rooms."
>
> (Group 2)

This transforming process illustrates an active learning process whereby students learn to reflect on their own learning experiences and develop the awareness of self-measurement.

8.4.2 Learning Processes in the PBL Environment

Before I started the empirical work, based on the knowledge about the PBL learning environment at AAU and experiences from previous research, I assumed that when studying engineering at this environment learning resources mainly came from three factors — doing project work in groups, attending lectures, and receiving supervision. With this assumption I started the empirical work.

Doing projects in groups

For the students at A&D, group work was the most difficult part to learn when studying in this environment. This was mainly because most of them had an upper-secondary school background from mathematical gymnasium, where they had few experiences in doing project work and group work. As discussed above, when they learned to reflect on learning, doing project work in groups became the major learning resources, from which they learned how to learn (learning methods), peer learning, management and organization, which were regarded as the most important things that have been learned when studying in the PBL environment at AAU.

Peer learning was regarded as the main difference between the PBL environment and the traditional lecture-based learning environment that they were used to. At the first semesters, students got confused and frustrated by sitting in the group room with 6 group members and spending lots of time and energy by talking and discussing. When they moved on to the later semesters, this way of learning became more and more valued. As discussed in one group:

> S1: “Now I think that it is a great way of learning. We don’t only learn from teachers but also learn from each other. I am not afraid to ask when I don’t understand because we are equal.”
>
> S2: “From my past experiences I know that I could learn from reading and writing, now I see that I can also learn from cooperation, and how to get along with other people so as to learn from them.”
>
> S3: “I learned to be open for other opinions. Through discussions we learn other people’s ways of learning … ”
>
> (Group 1)

As agreed in all of the 6 group discussions, group work was the most difficult thing to learn, but it was also the most valuable part with respect to the learning process in the PBL environment. In this aspect, according to students, the CLP course (that were provided in the first year aiming at helping students learn how to do projects in groups, as discussed in Chapter 2) was of great help to learn how to learn and how to reflect on learning when studying in this learning environment.

During the discussions, students in the female groups specially discussed how group work was of help with learning technology. As one female student said,

> "I am not good at technology, but I find it interesting to see how things work, so being in a group helps me to learning this. Besides, working in a group helps to keep people from dropping out. You don't lose your interest and drop out after you miss one month's education for example, because the others in the group will help to keep you up at the same level. At the same time, you have the responsibility for the group, and it is a duty to come to work everyday ... "
>
> (Nancy)

This experience was shared by different female students when talking about the strengths of studying in the form of group work. Having similar background, they felt equal with each other in general so that they could ask 'stupid' questions without feeling uncomfortable. Working in the open studio, they could get immediate help, because there would normally be some people around. Sometimes they could learn some technical skills just by passing by some groups and observing some people. In this way, their skills in computer and technology were improved rapidly through working in groups during the past years.

Studying in PBL environment, it is essential to make the group work function well in order to achieve the learning goals. Based on their experiences students talked about important factors that will help benefit from doing group work. As summarized in the following (also see Figure 8.1).

Management and organization were the most cited skills students reflected on learning from doing project in groups. According to them, how much they could learn in each project to great extent depended on how they could manage to work together by choosing specific methods, organizing group cooperation and managing the project. This process involves developing skills in plan-making, agreement-achievement, work-division, cooperation and communication.

Group	Factors	Quotations
1	Social atmosphere	We should make fun together in order to build up a nice atmosphere.
	Mutual respect	Everybody respects what others are doing, showing interesting and giving support.
	Attitudes	To be open-minded to different ways of working and thinking to be honest and frank with each other
2	Open-minded	We need be open up to different perspectives and methods and ways of working so that we can learn more.
	Mutual trust and respect	It is important that we respect each others' work, which is a way to learn new things We sometimes work at home, so we need trust each other. When people say that they work home, then trust them.
	Communication	We need tell others know what is going on in the group
	Combine work with life	It is very important that we have a balanced life, we work here and we all have different activities outside to recharge energy
3	Openness	It involves telling other people who I am, what I expect, and listening to others. To acknowledge the differences among people, and maybe I should not be the way I am all the time, I can have different new ways to think and look at things
	Attitudes	There will always be some problems in the group work, but I think that if we continue working on it, there will be lots of things to learn.
	Cooperation	It is one project work we are doing in our group rather than a mixture of different individual work. One important way we learn from the group work is through cooperating with each other.
4	Organization	We organized our time so that we both have time working here and also having time doing different things outside the university.
	Agreement and plan	It is important to agree on scheduling our time to make sure that we will finish things
	application of lectures to projects	It is very important that we use what we have learned from the lectures in the project. In this way it motivates us to go there and learn something. In our group we have agreed that we need attend lectures to get the knowledge.
5	Agreement	I think that it is important to talk about each others' ambition from the beginning, because people can have different ideas and expectations for the project. It is impossible to work together if there is no agreement.
	Mutual respect and trust	We need accept others' way of thinking and working. It is also important to trust each other.
	Supportive social atmosphere	I think that we are good at encouraging each other. When people get busy, it happens that they just choose something they are good at to make things easier. But we are quite good at keeping to what we agreed at the beginning, and be patient with each other and give time for each of us to try out things. Whoever is good at some specific area will be there and help others to manage this task.
	Organization	But in our group, we agreed to work together here most of the time, otherwise things will go loosely. It can be very frustrating and they end up with being very busy.
6	Structure	It is very important to have some ideas about how to work, and we are all very happy about the structure to do the project…
	Shared goal and plans	We like to know where we are going, It is impossible to work together if there is no agreement.
	Communication, compromise	We also need acknowledge the strength and weakness of our group members… we always make discussions in the group work. In the discussion, we should be able to say, 'I like this because…' rather than just because I like it… I think that's a very important part of in this way of learning in group work…

Fig. 8.1 Main factors for the function of group work.

Organization is of great importance to move on the project. At the beginning of the semester, students would talk about their expectations and learning objectives in the project as well as their strengths. Afterwards, they set up common learning goals and depicted an overview of the project schedule based on the human resources. In this way, they could both make the best of each other and meet the learning need of different group members. As discussed in one female group,

> S1: "we have different advantages so that we can make the best of each other."
>
> S2: "it is important that we talk about the weakness from the beginning, then we keep in mind that who wants to learn more in what areas."
>
> S3: "we try to bring some good experiences from past group work, at the beginning of our project work, we wrote the good experiences down and made them into a list, ... (show the list on the wall), ... it is not only about how to make the group work, also how to make the group report, how to illustrate different things ... "
>
> S4: "we made schedules, an overview one, how many weeks for the project, courses we are following, kind of big plan with big deadlines, we also have week schedules and daily schedules. We agreed to work together here everyday from 9:00 to 17:00, and later we will work longer and maybe also weekend ... "
>
> (Group 6)

In every group room there were two schedules (overview one and weekly one) hanging on the wall. This way of making plans was one of the skills that students learned in the CLP course in the first year. However, the ways of working on the daily basis were different from group to group. The three female groups have fixed schedule to work together in the project room, because in this way they could share information with each other, keep an overview of each others' progress and help each other when needed. The other three groups had different daily work schedules — sometimes working together in the group room, sometime working individually at home.

Flexibility was referred to as an important factor in the project organization. Schedules might have to be changed along the way according to the progress of the project and in case of emergency. One group talked about their flexible schedule which was different according to the stage of the project.

> S1: "It is important to work individually on different directions before we integrate all the different things. Generally, at the first stage, we collect information, which is normally done individually and we share knowledge with others."
>
> S2: "At the moment we do a lot of analyses and write a lot. We meet in the morning and do something together, and then we split up, writing something based on different tasks, then we read each others' writing before we meet again."
>
> (group 3)

The male group did not specifically mention their schedule, for whom, it was not necessary to write everything down on the paper as long as they discussed about things and make oral agreement.

Doing project work in groups involves group work and subgroup work as well as individual work. In general at the beginning of the project, a lot of discussions in groups were needed in order to choose topics and methods. When the overall plan was made, they would start to work in subgroups or individually. When asked how they divided the work, most students replied that they would try to find a balance between getting group work done and meeting the needs of individual learning goals. From the group perspective, they divided the work based on the human resources available and time schedules. From the individual perspective, they intended to choose what they were interested in, what they were not good at and what they would like to learn.

When speaking of management, the students talked about the importance of agreement in order to get work done as well as the difficulties in reaching agreement in their past experiences. At the first semesters, they spent a lot of energy and time arguing and discussing so as to reach consensus among 7 people. At this semester (the 6th semester), reaching agreement was not as difficult as before because firstly in most cases, they formed the group with people they expected to be able to work with, secondly; they had learned and developed different strategies based on past experiences. As one group talked about how they handle disagreement,

> Interviewer: "can you always agree on things?"
>
> All: "no, not all the time. It will be boring if we agree all the time."

> Interviewer: "how do you deal with that situation then?"
>
> S1: "We discuss about it. Actually we get inspirations and new ideas from the discussions. It is also a way to learn from each other. We listen to each other, and then find that maybe another person has a better idea than mine. Then we decide that what can make the whole situation better."
>
> S2: "We also need argument rather than only ideas, whether you like it or not."
>
> S3: "Yes, we need arguments to support the ideas, because we need to write them down in the report and we are going to have the exam. It is one project we are doing rather than just 6 small pieces."
>
> S4: "It takes lots of time having these discussions, especially at the beginning. But when we reach the end, things get easier and we finish things on time."
>
> Interviewer: "do you reach agreement at the end of the discussions?"
>
> All: "yes, we normally do."
>
> S5: "It involves compromise, sometimes some people need give up. Sometimes one person needs comprise to the rest when all the others agree, but sometimes one person might turn all the others around, if the person has a strong argument or proof, particularly in the technical part."
>
> (Group 1)

A common opinion shared by the interviewed groups was that 'it is impossible to get the project work done without being able to agree on things'. Thus they developed the strategy that they started the project with being open to each others' expectations. According to them, they made the agreement from the very beginning that it is important to agree on things, which would involve compromise and giving up some times. Achieving agreement and collaboration involves communication skills, which were addressed by the students as another important learning outcome in their experiences. Learning to communicate with each other was regarded as one of the most important factors that would increase responsibility and promote collaboration.

On the whole, a strong consciousness of the importance of learning from doing project in groups was witnessed through the group interviews in this study. When reaching the 6th semester, students gained strong awareness of setting up learning goals, exchanging explicit information about their strengths as well as their expectations. Social and communicative skills played an important role on building up a supportive and motivating atmosphere, which could lead to effective learning. Their emphasis on 'open and cooperative attitude' illustrated an active process in managing learning. Discussions in the group regarding how to relate information from lectures and textbook knowledge to the project work provides a context for students to understand technical contents in a deeper and more meaningful way. Peer learning through shared practice in the community is appreciated as an efficient learning strategy in terms of sharing information resources and getting inspiration. Good atmosphere and successful cooperation in the community of practice is recognized as motivation to get work done and to achieve the goal of learning. Working in groups can bring about mental supports as well as develop responsibilities in the learning process.

In general, a diversity of individual learning styles regarding learning preferences and ways of developing strategies of handling things was recognized by students as well as was witnessed in my research. However, some features standing out can also be generally recognized as gendered features. Among them, the most used words by women when speaking of their characteristics were 'structured', 'scheduled', 'well-organized', 'good at planning', 'communication'; whereas men often used words like 'logical', 'relaxing', 'better at technical contents', and so on. As discussed in one female group,

> S1: "I think that in our group, we are better at teaching and communicating."
>
> S2: "when I look back to my past experiences, I think that I have learned much more when I stay in girls' groups, because we teach each other. By teaching, I mean, explaining to others is a very good way for me to learn. So I have decided to work in girls' group in the following semesters. But I know that it is dangerous, because I know that I can learn from working

> in groups with boys, maybe it is just easier for me to learn when working in girls' group."
>
> S3: "I don't feel anything missing, because if we feel that we need anything, we can just go around to ask people nearby."
>
> (Group 4)

In the discussions of the female groups, students talked about their appreciation of the possibilities of working in both mixed groups and single-sex female groups. According to them, there were both advantages and disadvantages in either form of groups. However, this study environment made them confident in working with technology in girls' group. As mentioned in one group when discussing about differences between working in mixed groups and girls' groups,

> S1: "I think that it was a good idea to have some boys in the group in the first year. There were so many new programs we had to learn, which were quite hard, so it was easier to have boys around, because they have known them already and I could just go to ask. But now I have learned a lot, especially when working in girls' group."
>
> S2: "it is kind of natural for those boys to sit in front of computers. In many groups, the boys won't leave room for girls to do all those things, or guide the girls to learn it. They just do the tasks. Girls are better at making room for each other to learn things they would like to learn. We will say 'ok, if you want to learn it, try it and I will be here to help you.' Boys just don't do that. They think that it is natural that they do those things."
>
> S3: "yes, when you ask them some questions, they will just do it for you instead of telling you how to do it, because that is easier."
>
> S4: "but girls will be patient in teaching each other… "
>
> (Group 6)

Working in this environment provided chances for female students to learn technical things which would otherwise be regarded as men's task in group work. The general atmosphere as well as the physical convenience in the study environment made it easier for women to ask questions which otherwise might be regarded as stupid. It also provides chances for women to learn different skills to manage technical tasks, which would otherwise be occupied naturally by men.

Lectures

In Chapter 2 I have discussed the features of lectures in the PBL environment at AAU and the differences from those in traditional lecture-centred learning environment. A unique feature in A&D is that lecturers had very different backgrounds — they were generally from three different departments, Department of Architecture and Design, Department of Construction Engineering, Department of Mechanical Engineering, and external lecturers.

Based on the group interviews, how much students could learn from lectures were closely linked with two major aspects — curriculum design and the qualities of the lecturers. Concerning the influence of curriculum, there were three factors that matter to the learning outcome — the schedule arrangement of the lectures (which period of the semester and compression of lectures during a week or a day), the ways of conducting lectures (teaching methods) and the relevance of the lectures to the projects.

Normally all the lectures are arranged at the beginning of the semester, so that in the later part of the semester students can focus on doing project work. This way of arrangement was helpful to students since they could use what they have learned from lectures in the project work. However, in reality sometimes lectures could be squeezed together in some period due to lectures' availability or other reasons, which made it hard for students to handle.

> "It can be hard to learn if the lectures are too condensed. For a few times we had 8 hours' lectures the same day, which was so hard, and afterwards we had to do some project work. It is much better organized this semester, since we don't have lectures more than 4 hours a day so that we can keep a balance between lectures and project work."
>
> (Group 3)

As for the subjects of the courses provided each semester, the relevance to project was of great importance in terms of learning. As one group said,

> "When the lectures are not so relevant, it is difficult to use them, and it is difficult to learn. We attend the lectures, we listen, we understand or maybe not, and when we walk out of the lectures, we will forget if we don't have chances to use them."
>
> (Group 5)

According to the students, courses in general function in three ways: (1) to provide knowledge in certain areas that would be related to their project, (2) to introduce some research methods; (3) to provide some background knowledge. As long as they could be used somehow in their project, they could learn.

There were different ways of conducting lectures due to the lecturers' different backgrounds. Most students mentioned their appreciation of learning from lectures which would followed by exercises and assignments. As they said, they went to lectures, listening and taking notes, however, they could not really feel about learning until they got some excises or assignments to practice and discuss.

As many of them said, how much they could learn from lectures to great extents depended on the quality of the lecturer. When talking about the qualities of good lecturers, they generally mentioned two aspects: professional knowledge and the art of giving lectures. Some expectations to the lecturers can be summarized as following.

Firstly, personality was the most cited quality of a good lecture throughout the group discussions. According to the students, good qualities of a helpful lecturer include being 'active', 'energetic', 'catching', 'inspiring', 'devoted', 'responsible'. As an example described by students,

> "He is a very energetic person. He brings lots of things from real life, which makes the lecture catching. He tries to motivate us and keeps us active. When he gives assignment, we will be asked to do activities and then make public presentation. In this way our attention was kept all the time."
>
> (Group 4)

As most students agreed that a good lecturer could create a good atmosphere in the lecture room and could make boring and difficult contents interesting and easier to understand.

Secondly, a good lecture would provide chances for students to apply what they have learned to practice in real life.

> S1: "we had a lecture who told us some methods about video recording. We were given some assignments afterwards. We went to a café recording how a special kind of coffee was made. Afterwards, we analysed before we made a presentation explaining the procedures."
>
> S2: "In that way we learned through practice and explanation. It was very interesting and we learned a lot."
>
> S3: "And we never forget that afterwards. So if the lecturer can bring examples from real life, it will become more interesting."
>
> (Group 5)

In addition to this example, some students talked about their experiences of good lecturers who made contents easier to learn by integrating contents, relating them to each other and providing examples to illustrate.

Thirdly, a helpful lecturer knows pedagogy, and pays special attention to their presentation skills, for example, to provide handout beforehand, to have a clear structure and overview of the lecture at the start; to provide visualized presentation with colours, diagrams, pictures, etc.

In general, students agreed that professional knowledge of a lecturer was of importance, however, it is more essential that the lecturer had the awareness to place the students at the centre of education and try to know about the learning needs and expectation of diverse students.

Supervision

Studying in the PBL environment at AAU, every student group is provided at least one supervisor for each project. Supervisors play a role of facilitation to students' learning. The supervision system in A&D is rather unique due to the combination of two areas in the curriculum design. Students at 6th semester are supposed to do a project that cover areas of both designing and technical aspects (materials specifically). Therefore, two supervisors from different areas (designing and mechanical or construction engineering) were appointed to each group.

Based on data from the group discussions, learning from supervision takes place in different ways, depending on individual supervisors. In general, supervision functions in three ways: (1) providing knowledge inputs. Students ask questions and get information and instructions from supervision. (2) Providing reflection. Students are encouraged to reflect on their own work and experiences by the facilitating question like 'why', 'what', 'how' questions. (3) Providing comments. Students get feedbacks from supervisors based on what they have done and what they plan to do. All these three aspects could bring about inspiration of new ideas, provide a mirror for reflection and stimulate the occurrence of learning.

When talking about expectations from supervision, the most cited words students came out with were 'personality', 'engagement', and 'methods of supervision'. Some examples were provided with respect to good experiences of learning as following:

Firstly, when the supervisor had the readiness to make adjustment and provide different kinds of supervision to meet the different needs of students. Students have different expectations and demands for the supervisor when they reach different stage of the project process, which is also a changing process along the development of the project. As mentioned in one group,

> "At the beginning of the project, he needs to be more concept-oriented, to think about different concepts and layouts it involves in the project. At the end of the project, we might have very concrete questions. So the supervisor needs to be able to change the methods of supervision at different stages of the project."
>
> (Group 3)

Secondly, when the supervisor had the awareness as well as sensibilities of positive communication, which the students agreed that it should be the responsibility of both sides.

> "We realize that the problem is we know all the things in our project so that we also expect that the supervisor knows all about it. We need to let him know what is going on before we can get the right supervision. Sometimes it can bring about problems when we don't do it. But it is also a mutual way, because we also need to be able to let him know where we are in the project and what we need."
>
> (Group 4)

Problems would arise in the following situations, which brought about difficulty for students to benefit from supervision:

Firstly, when the supervisors were difficult to access. Sometimes supervisors could be too busy with other work so that they could not get immediate help when needed. Sometimes students could not get satisfactory feedbacks from supervisors because they did not have time reading their working papers.

Secondly, when miscommunication took place between the supervisor and students. During the group discussion, students talked about some situations when they had different expectation from their supervisors and when communication could not achieved. According to students, some supervisors were more interested in their own field and intended to guide the discussion towards their direction without paying attention to what students expected to gain from them.

In general, problems in supervision mainly arise from miscommunication, mismatch in terms of expectation and disagreement between two supervisors who have different backgrounds. For the students, miscommunication and different expectations from supervisors could turn into direct barriers to learning in this study environment. To use their words, 'sometimes we can be unlucky. When it (supervision) did not work, we had to get through everything on our own to get the project work done, which could be quite hard.'

Based on the experiences during the past years, students at the 6th semester had awareness of their learning needs as well as learning goals, based on which, they also established expectations to supervisors for more effective and meaningful learning (see Figure 8.2).

Group	Expectations	Quotation
1	Opinions, feedbacks, constructive critiques	'Sometimes it is not interesting if they just say yes to our work…' 'I don't mind get negative feedbacks as long as they can explain why, just to make us know that it will create troubles in this direction …'
	Methods	They will help us to move on rather than put their own agenda and ideas in your project.
	Engagement and commitment	They try to understand our ideas and give us different suggestions…
	Personality	They will have some kind of quality that make us feel that they can give something to the project, so that I want to listen to them and trust them.
2	Help me develop my way of working	It does not work for me just by telling me that you should do this and that. So for me, the good supervisor are the one who can help me experience, and guide me and provide me different options so that I can look back afterwards for better understanding.
	Guidance rather than answers	I don't want to know the exact way to do it, I want to get some general guidance about whether this is the right direction to go.
	Personality	If there is a strong personality, they will influence students and get people involved and give guidance.
	Constructive critiques	I would like to get critiques. I want to be shown when this direction does not work.
3	Have a clear understanding of the learning goals. Open-mindedness to different areas of knowledge awareness as well as sensibilities of positive communication	We realize that the problem is we know all the things in our project so that we also expect the supervisor know all about it. We need let him know what is going on before we can get the right supervision. Sometimes it can bring about problems when we don't do it.
	Capability to change in different situations	At the beginning of the project, he needs to be more concept-oriented, to think about different concepts and layout it involves in the project. At the end of the project, we might have very concrete questions.
	Engagement and commitments	In this project, we use Erik a lot, because he knows different things. If he doesn't know, he will talk to others or introduce us to others so that we can obtain the specific knowledge that we need.
4	Guidance rather than answers	It is good when they have different suggestions. But we don't want direct answers from them.
	Communication	It is better to have discussions than just one way talking.
5	Accessibility	We also expect to get immediate help when we have urgent needs. For example, when we get into troubles, we do need to get some help within a couple of days.
	Awareness of and intention to meet the expectation of students. Straight-forward opinions.	We call for a meeting because we want their opinions, after talking about our ideas; we would like to know whether it is good or bad, what their opinions are …
	Engagement	They don't have time to go into it to learn the things that we are doing. Their information sometimes comes from what we are telling them, so it could be because they don't know enough.
6	Methods rather than direct answers	I like the way when they tell us to try this or that. Instead of giving exact answers, they provide different options for us to find solutions.
	Engagement and interest	It is important for the supervisor to show interest in what the group is doing. Some supervisors have problems remembering different groups. I know that they have many groups to supervise, but it is kind of frustrating to explain everything once again each time.
	Awareness of and intention to meet the expectation of students	They have their ideas about what we should do in our project but do not listen to what we want to do.

Fig. 8.2 Expectations to supervisors.

Along the years, confronting mismatch in terms of communication and expectation with supervisors, these students developed different strategies to promote positive communication with supervisors.

> "I think that it is a good idea to have an agreement with the supervisor from the beginning of the project, but it is not simply just by a piece of paper (agreement contract with supervisors). We need find some ways to make it work."
>
> (Lee)

With respect to promoting general communication with supervisors, most of the students agreed that it should be a mutual responsibility for both sides.

> "We agree that it should be our responsibility for the communication (with supervisors) in order to get response from them. We should tell them things about what we have been doing and ask for response. Sometimes maybe we are not so good at it."
>
> (Group 6)

Some other groups also mentioned that they made agreement on initiating the improvement of communication with supervisors, and then they arranged meetings to talk about their expectations and learning needs.

In the light of the communication in supervision meetings, students also developed some strategies in order to get help. According to them, they tried to arrange things beforehand, for example, to remind the supervisor by giving schedules of what they had been doing since last meeting and the agenda for the meeting and their expectation regarding comments or help they would like to receive. As the examples some students provided,

> S1: "We can bring him to the right direction, for example, by saying 'shall we go to question no 2? Or shall we move on the discussion of next point." Otherwise it can be very frustrating if we won't get anything out of the meeting.
>
> S2: "I think that it is also a good way to learn by asking the supervisors why questions. Why do you think about it

> that way? Why do you put forward to these suggestions or comment?"
>
> (Group 1)

The process of developing strategies is also a recognized learning process reflected by the students. Most of them enjoyed this process by being active and initiative, because according to them, it turned out to be a process of self-motivation and self-management.

Social activities

Both interview and observation data in A&D showed that a positive social atmosphere is especially important for a supportive learning environment. The benefit from this was especially reflected since the 3rd semester when students became more experienced and familiar with the PBL learning environment and when they moved from the separate project rooms and the open studios.

Through doing projects in groups in each semester, students engaged in joint activities and established shared practice both inside and outside of their project group. In each group work, they attended lectures together, did exercises, met supervisors, conducted group discussions and arguments, wrote reports and took oral exams together. Working in studios, there were lots of chances to share information with other groups, to feel that they belong to a big community, to get help in time when needed, to have fun together during break time, and so on. The social atmosphere and social interaction through both formal and informal activities play a supportive role in the learning process.

A positive learning atmosphere is appreciated by students of both genders regarding seeking the meanings of learning. As male students in group 2 reflected,

> "In the group, there is some kind of chemistry that influences each other. If I come late in the morning, or if I need leave earlier in the afternoon, I might influence the other group members. I need explain to them that I do have something else urgent to do and let them know that it is not because I don't want to work as much as they do. It is important to be communicative and to respect each other. A bad circle will come out if one does not do the job and the others get annoyed

> and then nobody wants to get the job done. There is a kind of responsibility and we are sort of dependent on each other, like in a marriage."
>
> (Group 2)

A supportive learning atmosphere in project and group work also functions as a support for female students to learning technology, as the female students in group 4 agreed, 'working in groups we get mental support from each other, it is also a responsibility so that we won't drop out easily.'

All the groups agreed on the importance of building up a constructive social atmosphere in the group work so that everybody can feel comfortable speaking out their minds, feeling free to asking questions and sharing opinions with each other. It also leads to self-improvement and personal growth, as one student reflected, 'When I look back at these years' experiences, I said to myself that oh, I am able to communicate with so many different people and handle different situations!' The personal growth as well as the maturity that had been achieved through the learning experiences in group work was generally appreciated by many students.

In general, these students had more comments and reflections on group work and supervision than on the lectures. However, all of them said that the gender of the lecturer would not matter in relation to their learning outcome. Throughout the group interviews, there were no visible differences between male and female students regarding their interactions with and expectations towards lecturers and supervisors.

8.4.3 Learning Processes in a Semester-long Project Work

In this section, a report of the learning process of a semester-long project in one student group is presented, which followed the chorological development (See Figure 8.2). This aims to (1) provide a portray of the process of conducting a project studying in PBL, AAU; (2) confirm, compare with and make up information to the information provided in the previous text, which is based on group interview data.

From the theoretical perspective of the project-organized learning principle, students are expected to form the project group based on their interests in a research topic. However, in practice, it is more common for students to form the group before they reach the agreement on a topic. According to the majority of students, it is more important to be in a group with people whom they would like to work with. For them, on the basis of a communicative and supportive social atmosphere, it would not be difficult to find interesting topics to work on.

Tine, Arthur and Mike worked together in the same group in the autumn semester, 2003. After finishing the previous project, once when Tine and Mike were hanging out, they talked about working together on a new project, because they had a good experience cooperating with each other last semester. They discussed about some interesting topics and decided to work on that. They talked about it with Arthur, another group member whom they worked with in the last project. Arthur responded positively and showed immediate interest. Once the idea of working together was settled, they started to talk about it in public (during gathering or meeting with other students) because they needed one or two more people to form a group. Sara and Michael showed interest when they heard the proposal of the formation of this group.

Thus, this group was formed informally during January without clear ideas on the topic. Since nobody showed indication of changing mind before the starting day of the new semester, they registered their group formation in the group formation section of the introductory activities on February 1st.

February

The 1st week

February 1st, 2004, new semester started. After the official formation of the group, they went to their project room for the first meeting. It was an around 10 square meters' cubicle in the big open studio divided by partitions. At this moment, the room was rather plain, with only three desks, 6 wheeled-chairs, two boards, and some book shelves.

The first meeting was mainly focused on introduction and getting prepared for the start of the new project. In turn, they introduced their experiences in the past projects, their strengths and weaknesses, their expectations of learning in the coming project. Afterwards, they discussed about practical issues on planning and organization of the group work. It did not take long time for them to reach some provisional agreements with respect to working hours and organization of work. They agreed on a trial of a short-term plan (the first months): (1) to divide the work into individual or subgroup tasks; (2) to work

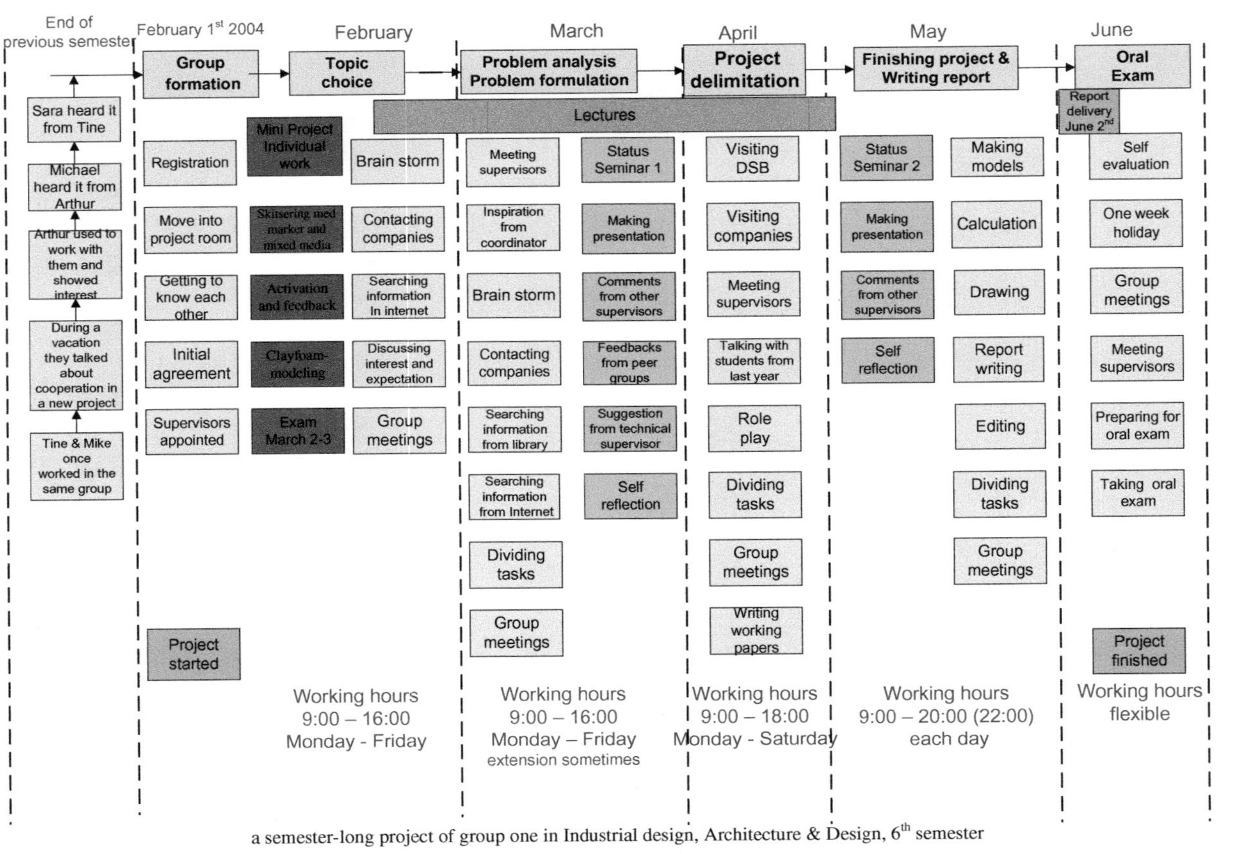

Fig. 8.3 The process of a semester-long project work by Group 1 in spring 2004.

in the project room everyday from 9:00–16:00 (except for the lecture time) on their project and do other tasks individually outside of this time period.

Then they moved to the discussion of the topic of the project. By then, they only had some general ideas based on their shared interest with respect to the general direction, which was based on their shared interest in doing a project in medical area. Arthur suggested using a brainstorm to collect some concrete ideas, though it did not lead to any specific area in the next half an hour. Thus Tine suggested that they could try to contact different medical care related companies to get some inspiration or even agreement on some projects. They agreed that it could be the tasks for the following days. After the meeting, they explained to me what happened during the meeting. They regarded it as a good start since they reached agreements on organization and planning. They were not worried about finding an interesting topic, because for them, the first important thing at the beginning was to build up a communicative social atmosphere in the group. As Michael explained to me, 'I don't think that any group will decide the topic before they form the group, because when you decide whom you want to work with, you can always choose what you want to work on.'

In the following days, they used resources of previous knowledge, the internet and a telephone book to collect a list of relevant companies before they contacted them either by phone, email or by visiting. They did not find any satisfactory results because either most companies did not have projects available or the options were not attractive to them.

By the end of this week, they had a discussion based on the situation, and agreed that it might be a good idea to switch to another area than medical care instruments. In the new round of brainstorm, two catching words came out: handicapped and train. Doing something for handicapped people was inspired by one of the proposed areas from the coordinator, Phil, in the introductory course. The train station provided a context for concrete ideas. During the discussion, everybody showed interest in the idea of the train station; because it was one of the most commonly visited public places.

During the brainstorm process, they came out with different ideas and possibilities, some of them were more joking rather than serious ideas. This discussion of train and transportation tools was associated with various fancy technologies from movies, science fictions or from imaginations. Some ideas led to different jokes, which brought about laughter along the way. As they said, studying in this way brought a lot of fun, which combined work and entertainment, especially at the beginning of the project.

Another round of brainstorm led to a narrower focus — something useful for handicapped people in the train stations. It did not take long before they reached a shining point that made everybody excited: how to help handicapped people get on the train easier — a lift. Once this idea came into being, they made a list of 'does' (tasks) for the next week. The list covered two main areas:

1. to contact DSB (Danish railway company) for information on (1) what procedures handicapped need to follow in the process of entering a train station and getting on the train; (2) what facilities were available for assisting the process.
2. to contact some companies that produce products for the handicapped to see what were available and what were needed.

Meanwhile, during the week, they were appointed two supervisors — Nis, the main supervisor, who is from A&D and has a background of industrial design; Eric, the technical supervisor, who is from Department of Mechanical Engineering and has a background of mechanical engineering. Generally, the main supervisor is responsible for the overall project design and process. The technical supervisor is expected to assist the project work with concrete help in perspective of technical and engineering parts.

During the first week, some groups initiated the first meeting with their main supervisors for inspiration and recommendation. Group 1 chose to wait for the first supervision meeting until they had some brief ideas of the topic so that they could have more efficient discussions. Sometimes when Nis was around having meetings with other groups, he would pass by this group to say hello and to have some informal talks to see how they were doing.

At 4 o'clock in the afternoon, they finished the work for the week and left for weekend. They agreed that it was important to have a life outside of university, where they could get energy for work.

The 2nd to 4th week

From the second week, the mini (also named minor) project started, which are made up of three one-week-long sections. The minor project was concentrated and demanded lots of time for individual work. The main project was laid off temporarily in most of the groups. Students went to lectures in the morning and did assignment by individual work in the afternoon.

There was not much progress in the main project in this group either. They sent emails to asking for information from DSB and a company which produce products for handicapped people. They got a reply from the company, showing interest in their proposed project and suggested more contact with clear and detailed information. They informed the company that they would like to come back to visit them either by phone or by visiting at the beginning of March after the exam of their mini project. They did not get response from DSB.

March

After the mini project exams, all the students moved their focus to the main project. Most of the lectures were provided in March and April. In this period, they attended lectures 4–6 times a week (see table semester schedule). Lecture time varied from 8:30–12:00 to 12:30–15:30.

From this period, students in group 1 decided to spend more time on some research in order to agree on a topic for their main project. They worked together from 9:00 to 16:00 in the project room from Monday to Friday except lecture time, and worked individually in the evenings and some time during weekend.

March 8th the first formal supervision meeting

The first official meeting with the main supervisor, Nis, was arranged on March 8th. Before this meeting, they had different informal talks about the brief ideas of the project. In this prepared and planned meeting, students presented their ideas, status and expectation about the project. They organized all the papers, documents and brochures about the problems which handicapped people faced when getting on the train and assistances that had been provided by then. Arthur was responsible for the initiation, arrangement and management for this meeting. So he showed Nis the documents and started the meeting with introduction to their changing process from medical care at hospital to a lift for handicapped people at train stations. Sara was responsible for note taking at the meeting. A communication map can be seen as Figure 8.4.

> Arthur: if handicapped people want to travel by train, they have to arrange it 2 days in advance. Then the train station will prepare the needed facilities for them. This makes travelling

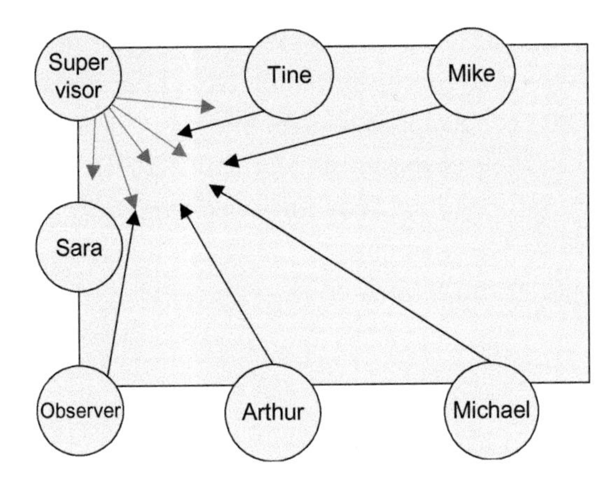

Fig. 8.4 Communication map of the supervision meeting at the beginning of the project (group 1).

complicated and inconvenient for them. So design something to reduce their need for help.

Nis: it sounds interesting, so what information have you got up to now?

Sara: not very satisfactory actually, we contacted DSB and were told that they were busy at the moment and would call us back, but we haven't heard from them yet.

Nis: oh, it is a pity. Have you thought about going there to check the situation?

Michael: yes, we actually plan to go to the train station on Thursday.

Mike: we have contacted another company which produces some products for handicapped people. It is a small company, and they were quite helpful.

Tine: those were the places we had contacted, how do you think about that? Do you have any other suggestions or ideas?

Nis: If you contact Danish Handicapped Society, maybe they know about some possibilities to see what they think about our ideas and what information we can get from them.

> Tine: that sounds like a good idea. We would like to broader the use of the lift so that it can be used in different contexts rather than only train stations.
>
> Nis: how about other facilities in other companies or other countries? There might be a chance to see what are available in other contexts and bring about some inspirations. In some countries there might be a better overview ...

Then they discussed how handicapped people managed things in their daily life. Nis suggested that they could use a scenario of how handicapped people get on the train at the station to see what the problem would be.

The next topic on the agenda was the working paper about problem analysis. Arthur talked about the group's ideas about that they were going to write based on the current stage. Briefly, Nis gave a positive comment on these ideas before he expressed his expectation about the contents of the paper, and how they could write it. Specifically he stressed the importance of the consideration of the users' needs in the process of planning the project. At last, he provided some references to some literature about expert users.

After the one-hour-long meeting, the students had a discussion and decided to pay a visit to the company which produce wheelchairs for handicapped people because they had been invited. They also adjusted their 'to do' list based on Nis' suggestions.

March 25th the third official meeting

Meetings with supervisors fell into different types based on time, contents and needs. Official ones were more planned and structured, which lasted one to one and a half hours. Informal talks were another agreed way to get supervision in the situations that Nis passed by to see how they were doing, or students asked for 10–20 minutes meeting for questions that would expect short and quick answers.

This was a planned meeting but without preparation. The main purpose of this meeting was to get inputs. Students in group 1 became frustrated when they had no reply from DSB for the information they asked for. Without this basic information, most of their analysis was just based on imagination without confirmation. They asked for a meeting with Nis in order to get some inputs.

When Nis heard their explanation of the situation, he suggested another scenario discussion to see what the involved factors were. From this round of the discussion, some new issues came out such as weather conditions and environmental considerations. They discussed about using case methodology and possible solutions based on different options. With the help of Nis, they agreed that some options involve redesign of the train, which demands too much information.

From this meeting, they got an overview based on different situations, but again, they were still dependent on some important information from DSB.

March 29th

It was the first status seminar where all the 8 project groups were supposed to make a concept presentation about the basic concepts which they employed and the current status of the project. All the supervisors and lectures involved in this semester curriculum were expected to be present. The purpose of this one-day seminar was to provide a chance for reflection and for comments from their own supervisors as well as outsiders like other teaching staff and peer students.

Five teaching staff participated in this seminar, two main supervisors, Phil and Nis, and three technical supervisors from Mechanical Engineering and Construction Engineering respectively. They were all involved in supervision work and some of them gave lectures at this semester as well.

Every group had 45 minutes for presentation, answering questions, discussion and feedback. Based on experiences from the last three years, students were not nervous about making plenum presentation. They were experienced in designing and laying out their power point presentation. Students in group 1 prepared for this presentation after the supervision meeting. One was responsible for the general layout of PowerPoint slides; two were responsible for drawings; two of them were responsible for preparing background and concepts knowledge for the problem analysis. They did a brief rehearsal beforehand during lunch time and all of them took a portion of the presentation when their turn came in the afternoon.

Their presentation lasted about 25 minutes, including clarifying questions along the way. Afterwards students from other groups asked questions and gave comments as well as suggestions. This is followed by feedbacks from some of the teaching staff.

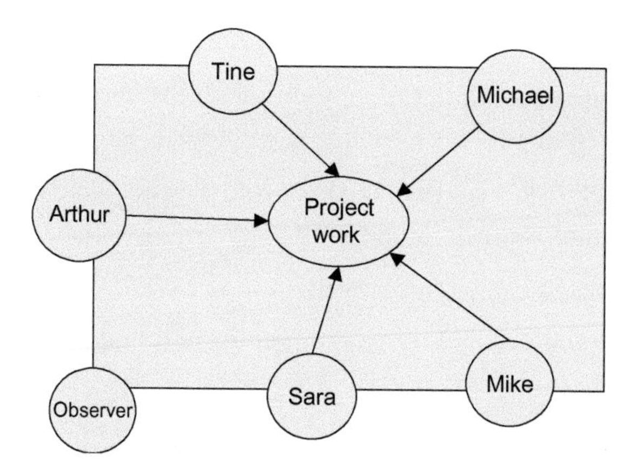

Fig. 8.5 Communication map of the group meetings (group 1).

The next day, in the group meeting, students from Group 1 talked about the inputs they gained from the seminar. As extracted from their discussion (Figure 8.5 shows the communication in the meeting):

> Sara: I think that it was interesting to see others' work, and to have some knowledge about their products…
>
> Mike: yes, some of them brought new perspectives to what we are doing.
>
> Tine: yes, it is nice to have some constructive feedbacks, and quite some of them were good ideas, I think.
>
> Michael: yes, it was interesting to observe other's group presentation. I think that it was a good idea that all of us participated in this presentation, rather than only sending representatives…
>
> Sara: yes, I think that I liked the slides design of group 6, it is fun to use cartoon.
>
> Michael: I am not sure I agree with that, it did not look very serious…
>
> Arthur: it was fine for the group with those girls… I am not sure that I will do it… but I am thinking that we could have made some of our slides more understandable…

Tine: how do you think about the comments from Phil?

Mike: it is useful, but more focused on the design part. We need to find a way to combine it with the technical parts.

Michael: Eric's suggestion about role play could be interesting. I think that we should try it.

Tine: yes, it can be fun as well.

Arthur: yes, we can try it and then we will find out whether it will fit in.

In order to sum up, they listed different aspects that they had learned from: (1) they got to know more about the project work of other groups, which was interesting and inspiring. (2) Constructive questions from peer students provided chances for self-reflection about their own project as well as ways of presentation. (3) They also shared some information with another group who worked on some facilities in medical care at hospitals. (4) It was a good opportunity for practicing presentation skills by using PowerPoint. (3) New perspectives were brought about from comments from teaching staff. Especially, they got a very inspiriting idea from their technical supervisor Eric that they could use a role play as a strategy to get information and ideas.

Up to this period, the project went on well, though a little bit slower than planned. But it was not surprising for these experienced students. They were satisfied with the supervision as well. From Nis, they got different new perspectives on the overall design of the project, some of which were something they never thought about before. Nis helped them formulate the concepts and analyzing things from different perspectives: if they do this, it would lead to some certain situation, it they go another direction, it might lead to some problems.

They also arranged a meeting with the technical supervisor Eric, who was very helpful and resourceful in providing information. He introduced different literature with theories in the technical design part. In a couple of days, he came back with some referred books and some marked the relevant chapters to the students for reading. Later on he passed by a couple of times to see the progress of the project work and explained the technical things that could be difficult for them to understand. Considering that it was a new area for them, he tried to explain things in a clear way with examples relating to design areas in order to help them obtain a basic understanding. His interest in and engagement to this project motivated the students to get involved in the work.

In this month, they had a lot of fun by staying together in the project room — having lunch together, chatting in the break time and discussions about project work. Once a while, they went out to have some ice cream or beer together. Nevertheless, they agreed on the importance of having a life outside of the university. Thus, they tried to keep the working hours in the group to the plan (9:00 to 16:00 except for lecture time), so that they can have time for individual tasks, reading work for lectures, and spare time to spend with friends or sports. However, in some days the work hours were prolonged when they had more exciting ideas about the project work, or when the discussion could not lead to the expected result on time. Sometimes they brought some tasks to do at home during weekend.

April

In the spring semester, April is normally a transition month in which lectures reach the end and project work goes into depth. According to many students, this month was challenging because of the double tasks — managing both attending lectures and proceeding project work. Especially the key factor was the application of lecture knowledge to the project work.

On April 29th, the coordinator Phil organized a special activity for students before the second status seminar. It was an extra lecture aiming at helping students applying knowledge that were gained from lectures to their project work. According to Phil, this was a new experience for him as well. Based on his past experiences, at this stage of the project work, students tended to forget what they had learned in the mini project and in the lectures (which were mostly provided at the beginning of the semesters. So he searched a lot of pictures from the internet and showed them to the students. He used two screens in the presentation to show similar design at different time and cultural backgrounds. The purpose of this activity was to remind the students that they should relate the theme of design to their project, which covered different concerns and cultural meanings. The idea of introducing this method at this moment was not to tell students 'what they should do', but rather to inspire them and help them open up their minds by introducing what other people did according to Phil.

Afterwards, students were asked to do research on the internet by using keywords. Students from Group 1 chose 10 words which were associated with their project before they did the search on the internet.

This month, the project work of Group 1 proceeded unsatisfactorily. They went into stress because of the lack of relevant information about the train station from DSB. At the end of the month they decided not to wait any longer but instead they employed the strategy of the shifting the focus from the train station to the lift itself.

During this month, they only had a couple of informal meetings with Nis. Facing slow progress, they would like to have some direct instruction or suggestions that would lead to some more forward movement of the project work. They were a little bit disappointed with Nis when he only showed understanding of their frustration and stress rather than providing some effective strategies. With the help of the technical supervisor Eric, they developed some strategies to put more weights on the technical part of the lift.

Just before this activity they were contacted by people from DSB and got the information that they had asked for two months before. They felt that they were put into dilemma situation again: to come back to the original plan or to go ahead with the modified plan. Getting pressured, they were not very devoted to this activity of searching key words on the internet. However, they chose to do it briefly since it might be fruitful somehow.

They agreed not to use too much time on this activity, so they came back to the project room for meeting. Different opinions were held among the group members: Sara and Michael suggested that they should go back to the original plan since it was based on their shared interest. However, Tine and Arthur proposed an easy and safe strategy, that is, to adjust their goal of the project to finishing rather than being too ambitious since the due time was approaching. The argument of time limitation convinced the rest of the group members, which led to the final agreement that they would keep the modified plan.

In this month, working hours were prolonged little by little since they had more and more tasks for the projects. They also spent more time than expected on discussing different strategies to move on the project and modifying schedules and plans. Therefore, they agreed to work from 9:00–17:00 or 18:00 some weekdays, and some time of the Saturdays.

May

May is the busiest month for students in the spring semester. Normally all the lectures have finished by then; the project work approaches the last period when students are busy with finishing the report writing before the due time, the beginning of June. Since the beginning of May, most groups prolong their working hours step by step from 9:00 to 18:00 to 19:00, 20:00, until the last two weeks 9:00–21:00 from Monday to Saturday (some groups worked 9:00–22:00, some groups worked 7 days a week).

May 4th second status seminar

The coordinator Phil organized the second status seminar, which aimed at reminding the students of the overall plan and pushing those groups who were behind schedule. During this seminar, most groups have reached the end of their project. At this stage, students had gone through most of the research process and became the 'expert' of the project. It was difficult for supervisors of other groups and peer students to comment on the project work. Thus this seminar mostly functioned as a milestone for each group to regulate and reflect on their own work. As the students from Group 1 said, 'it is difficult to get direct inputs from outside when we are at this stage of our project, because we know the most about our own work. So this seminar works as a way to push us to organize our work and check it out where we are, how far we need to go, and such things.'

May 18th supervision meeting and group meeting

In this month, the project work in group 1 proceeded well after they developed strategies based on the modified goals. A meeting with Nis was arranged on May 18th when the brief model of the lift was worked out, which was an important milestone of the project. Tine was responsible for organizing this meeting. She gave a review of what had been done with respect to the progress of the project. Afterwards, she asked Nis to comment on their work, 'this is what we have been doing recently, based on what I have described, what do you think about our project work? At this stage, we would like to get different kinds of comments. So please feel free to give us your critiques.' Nis gave quite positive comments on the model and analyzed some possible problems it would lead to. Based on these, he put forward some suggestions for modifying the design.

Picture 8.1 The model made by students in Group 1, A&D.

After Nis left, they had a group meeting discussing strategies based on Nis' comments and suggestions. Tine suggested that everyone of them made drawings of different parts of the lift based on Nis' suggestions before discussion. The rest agreed. They were commenting on the suggestions while making the drawing. They talked about in which way and to what extent Nis' suggestions could fit into their situation.

There were not so many arguments in this meeting. Consensus was easier to reach when the due time of report delivery approached and the goal of finishing on time was shared. They agreed that most of the analysis and suggestions were reasonable, though they disagreed on some of them. However, based on the time limitation, they could not make all the suggested modification. They decided to apply some of them and give up the ones that would lead to big changes. It was more important to focus on one structure and make reasonable arguments for their choice of methods and details in order to finish the project in 2 weeks.

During the discussion, Arthur was making notes on the blackboard by writing key words and making a task list for the next two weeks.

— Time plan
— Interview with people at the train station
— Layout
— Sources
— Pictures of process
— Material choice
— Battery, size, operation time, source
— Ergonomics, overview of weight distribution
— Engine
— Station analyse
— Station + interface
— Handle on the backside of the lift
— Diagram techniques
— Interface PDA (feedback)
— Scissors

At the beginning of the project, they mostly chose the tasks based on their personal interest and learning goals in certain areas that they were not good at. When reaching the end of the project, the distribution of tasks was sometimes based on practical situation or strengths of the individuals. At this stage, students in group 1 were trying to find practical methods in order to get the project work done on time.

May 26th last supervision meeting

Progress of the project could be witnessed from the sets of posters and drawings hanging around the project room. Students were spending more and more time in this room. The space was made fully used for their work — windows and walls with posters hanging; the open area with the models standing; documents and papers piled up on the desks side by side with plates, knives and forks; food package and bottles of drinks stocked in one shelves.

In the last meeting with Nis, they worked out a list of things that had been missed: requirements with focus on what handicapped people need, and strategies on how to build up a communication channel between the train driver and the passenger. Some questions were discussed in perspective of the weight of the lift and security considerations. The meeting was precise and short. Some of them were having lunch while discussing. As can be seen from the communication map (Figure 8.6), at this stage, supervisor meetings had been

Picture 8.2 Project planning in group 1 at A&D.

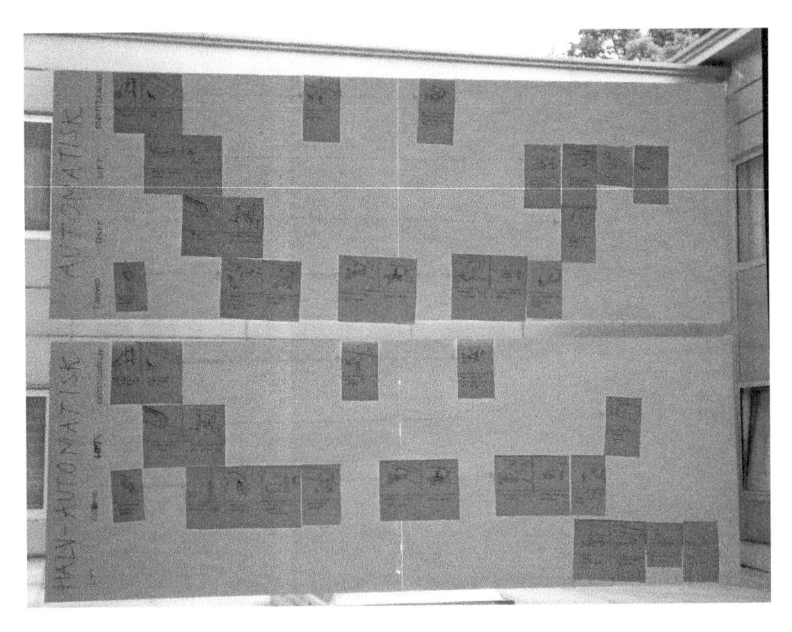

Picture 8.3 Project outcome of group 1 A&D.

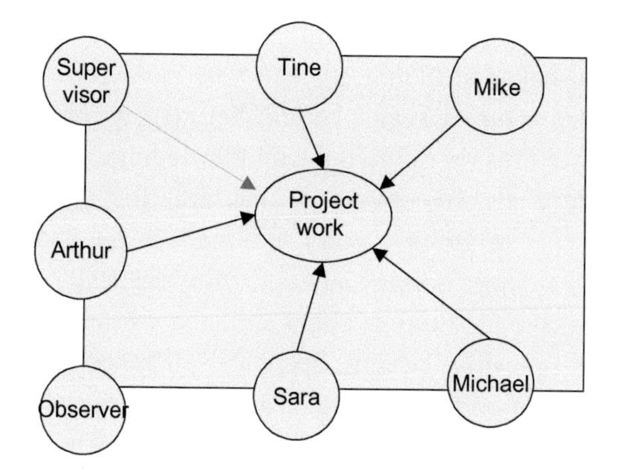

Fig. 8.6 Communication map of the supervision meeting at the end of the project (group 1).

changed from students-supervisor conversation model with focus on inputs provision from the supervisor to the group discussion conversation model with focus on discussion about the project. Expectations from supervisors were changed from providing information to giving comments and critiques.

The working hours had been increased step and step in May. In the last two weeks of May, students in most groups worked from 9:00 to 21:00 (and even longer) seven days of the week. This was a period with concentrated work (a few sleepless days in the last week) without social life, according to them, no sports, no bars, no contact with friends. However, they could live with it for this period because they were motivated by the clear goal of finishing the project and fulfilling their tasks of the design.

Students in group 1 were rather scheduled. They proceeded well in the last month and by the time of the last supervision meeting, they finished the majority part of writing the reporting. In the last week, they finished the whole report and worked on modification based on the critiques from Nis. After the last round of editing, the report was finally printed and delivered on May 31st, one day before the due time of June 1st. So they were released when a few groups were still at hard work in the last day.

June

After the delivery of reports, students in group 1 had a few relaxing days. Most of them chose to make up the sleep that had been missed in the last month.

Their oral exam for the project was arranged on June 26th, which would be the last group. It was not an ideal time since they had to wait for another 4 weeks. Thus they decided to take a week's holiday before they came back to work on the preparation for the oral exam in the middle of June.

This semester-long project came to the end after their examination on June 26th. Students in group 1 got a fine mark in the exam.[8] Afterwards, they started their summer holiday. In September, during the study trip for the intercultural workshop in Beijing,[9] I got a chance to have more informal talks with students in Industrial Design. Specifically I asked students in group 1 about their reflection on this project. The three of them who participated in this trip said that it was not as good as they expected due to the lack of needed information at the beginning. However, the most important thing they learned from this project was how to handle unexpected situations in order to go ahead with the project and finish it based on what were available. They also expressed the plan of doing something new in the next project.

To summarize, in this section, both interview and observation data portrayed the learning processes of studying in the PBL environment in the program of Industrial Design at A&D. The assumption that learning resources in the PBL Aalborg Model mainly come from three parts (project work in groups, lectures and supervision) can be confirmed. However, the research findings exemplify that there are more resources than these three major ones, for example, social activities. This process of learning from diverse and multiple resources illustrates an active learning process.

8.5 Professional Identity Development at A&D

Studying engineering in the PBL environment, students in A&D learned certain methods, developed strategies, participated in joint activities and established shared practices in the process of carrying out projects in groups. This

[8] For the consideration of confidentiality, I do not report their mark.

[9] At A&D, there is a study tour for students to travel aboard at the 6th semester. In 2004, the tour was postponed to September and was conducted in Beijing, China for an intercultural workshop (with a theme of designing some product for the Olympic Games in 2008). In the course of my fieldwork in spring, I participated in the organization of this activity, helping them with some practical issues. This helped my research by building up closer relationships with the students. The company with them in the workshop also provided some chances for informal talks about their reflection on the projects in the spring semester.

is also a process whereby they developed a professional identity. The professional identity development in the leaning processes of students at A&D can be witnessed from the ways they reflect on their learning experiences, different values they received from teaching staff as well as the ways they set up learning goals and develop their own learning strategies.

8.5.1 Critical Reflection on Learning and Self Identification

As has been discussed before, in the learning processes, students developed skills in reflecting on their experiences. This reflection helped them see the meanings of learning and appreciate the feeling of meaningfulness in the learning processes. In addition, my research also observed critical thinking and reflection of these students, based on their comparison between themselves and students from other engineering programs and other universities. This reflection can also be seen as a way for them to identify themselves professionally. The main aspects can be summarized as following.

(1) Basic skills. More than half of the groups mentioned that one of the main differences regarding the way of studying between AAU and other universities (from their knowledge about their friends' experiences) lay in how much theoretical knowledge students learned as basic skills. Some groups agreed that it could be a good idea if they spend the first semester intensively on different new subjects so as to build up a solid basis for project work at later semesters.

> S1: we discussed a lot about this. We think that it might be a good idea that when we started we spent the first year intensively on different subjects, like high schools.
>
> I; which year do you prefer to do that?
>
> S2: maybe the first year. But it was also important to learn how to do projects as well. So maybe just a semester.
>
> Interviewer: in your department you actually have a mini project each semester for individual work.

> S3: Yes, but it is at a lower level than expected. We still focus on the main project, and it was just 3–4 weeks. We could not spend more time reading things.
>
> (Group 3)

(2) Individual learning and working abilities. Most students mentioned the paradox in terms of a balance between collaborative learning and individual learning. On one hand, they appreciated learning from doing group work; on the other hand, they felt something missing with respect to individual learning. It takes a lot of time and energy to work in a group, therefore, they have little time for reading books to prepare for the lectures,

> "It would be better if I had more time and energy to sit down for more reading and thinking in depth about what I have been doing. I feel that I have been active all the time here, which takes quite a lot of energy. When I go home, I mostly just want to relax. It would be good if we have some time to be quiet and read."
>
> (Tine)

This aspect is also reflected in their consideration of improving creativity and learning the unique way of thinking as a designer.

> S1: "Working in groups definitely helped with inspiration for good ideas, but we are also individually different. Sometimes I feel that I want some space for myself."
>
> S2: "We are told that working with technology and working in companies involves lots of collaboration, that's why we need develop our skills to do group work. But I think that being designers also involves doing individual work ... "
>
> S3: "Yes, designers are supposed to create something new and unique. That's why we want to become designers."
>
> (Group 5)

(3) Diversity in terms of study methods. Most students reflected the way of learning in the PBL environment by comparing themselves with their friends or acquaintances who study in other learning environments. This reflection helped them see various advantages of this way of learning; however, they felt tired of and bored sometimes. Getting used to one fixed way of working can make study life easier; on the other hand, it might hinder the development of new skills and creativity. As reflected by one group,

> S1: "I have friends who study at different universities around the country. I can see that the ways we study are very different from each other. Most of them just go to lectures and then read a lot of things on their own, and then they go to exams. Here we do a lot of projects and are involved in many activities."
>
> S2: "Well, to be honest, sometimes we are tired of projects even though we know that in the long run, there are many advantages."
>
> S3: "Sometimes I feel that it is endless to study by doing projects, when this project eventually finished, next one came."
>
> S1: "From comparing with my friends, I can see lots of good points from the way we study here, but I am also thinking about how to develop different working methods rather than only one way of thinking."
>
> S4: "Yes, I agree, I am actually thinking about similar things. When we know that this is a good way of learning, then we tend to do it again and again, but then how can we develop new things, and how can we be creative..."
>
> (Group 4)

(4) Interdisciplinary learning, which was reflected through students' comparison between the methods of doing projects among different engineering programs. According to these students, project

work at A&D involves broad topics and interdisciplinary knowledge and information. As agreed by students from different groups, being designers, they need to find problems to solve rather than waiting to be given a topic by supervisors. Their project topics normally covered different aspects, which demanded searching specific knowledge of those different areas. As Martin talked about his opinions on the differences between industrial design and electronics based on his past experiences of studying electronics in technical gymnasium,

> "As designers, we do projects that cover broader areas. Every time when we do a project, we really need to go out to do different research. We need contact people outside of the university to know this and that in order to gain different knowledge. In the Electronics area, students go in depth with one focus but do not cover different big areas. For example, they solve problems in electronics but they won't go into the material part. Well, we do a lot, so it might be nice if we can cooperate with students in other programs."
>
> (Martin)

This comparison was specifically addressed in the discussion of group 2 (the male group). Three of them had a background of technical gymnasium, where technical knowledge was placed as the focus of education. All of them said that they enjoyed the experiences at A&D, which involved broader areas of knowledge compared with their past experiences in technical gymnasium. However, they mentioned the general lack of collaboration between different engineering programs. They expressed their wish of building up a network among students from different study programs so that they could 'share information and practice more challenging projects in order to achieve competences at a higher level'.

From what has been discussed, students at A&D reflected on both advantages and disadvantages of the studying methods in the PBL environment based on their three years' experiences. This reflection helped them see both sides of the coin as well as develop a sense of professional identification.

8.5.2 Different Value Resources from the Teaching Staff

Teaching staff play a role in establishing the values of the learning culture. However, this role at A&D brings about confusion to students because their lecturers and supervisors come from different professional backgrounds. The teaching staff in A&D are made up of experts from two different areas — engineering and architecture & designing[10] — who bring different expertise, experiences, and ideas about how to how to develop the culture of the department.

The differences between teaching staff from different backgrounds were clearly identified by students. This topic was discussed in all the groups. According to the majority of the students, they had very different opinions on the teaching objectives and teaching methods. Students often said that 'they come from different worlds'; 'they speak different languages'; 'they have different ideas and opinions' and 'they do things in very different ways'.

Based on the group discussions of students, the main differences between their lectures and supervisors who were engineers or designers/architects as well as the differences between the two disciplines of engineering and industrial designing/architecture can be summarized as following.

(1) Engineering is more products and solution-oriented; whereas designing and architecture is more process-oriented.
(2) Engineers are more logical and rational; whereas designers and architects are more creative and abstract.
(3) Lectures on engineering will start at 8:30,[11] and lecturers will arrive 5 minutes beforehand; whereas lecture on designing and architecture will start at 9:00, and lecturers often arrive 5 minutes late.[12]
(4) Supervisors from engineering are more focused on technical parts of the project. This is reflected from the questions they ask, which would normally expect some precise, fixed, single-truth based and correct answers. Supervisors from designing and architecture are

[10]There were different ways for the teaching staff at A&D to identify themselves professionally. Some were designers and some were architects. In my reports, I put these two into one group as a way to distinguish them from engineers.

[11]This is the schedule in A&D. As reported in Chapter 7, in the hard-core engineering programmes, like EE, the normal daily schedule starts at 8:15.

[12]This might be over exaggerating, but it is a visible difference agreed by most of the students.

> more focused on the design process and artistic aspects. It is difficult to answer their questions with facts or from a single perspective, but rather conducting discussions and building up arguments. There is no belief in fixed and right/wrong answers, and students are expected to analyze things from different perspectives.

In this way, students often get conflicting responses, comments and suggestions from their lectures or supervisors who were from these two different backgrounds. As discussed in most of the groups, sometimes it was frustrating when the teaching staff (especially supervisors from two areas) could not agree on the criteria in the evaluation of their project work. Based on this, the learning process in A&D also involves a process of finding out how to compromise and integrate opinions from different supervisors. They tried to combine both of the aspects through their work, because for them this was the reason for their choice of this study (that is, to learn things from two different areas). When speaking of strategies, some groups mentioned the importance of having clear ideas of their own learning goals and the aims of the projects. Some groups learned to 'figure out what is the most important and useful parts to the projects and say no to some of the suggestions from supervisors'. When facing serious disagreement between the two supervisors, some groups chose to arrange meetings with the presence of both sides for discussions in order to reach consensus at certain levels.

Therefore, most of the groups expressed their learning goal of 'exploring their own ways of doing the project.' This idea developed into a shared belief among students from their learning experiences, that is, to learn from different approaches and to develop their own approach. As many of them said 'we are developing new ways of doing things and this should be the idea of this study program'.

8.5.3 A New Type of Engineers

This active learning process of students in A&D not only led to their participation into the establishment of the new learning culture, but also the construction of a new type of engineers.

There had been plenty of debates regarding how to define the professions of students in A&D. Officially, it took three years before this program was accepted as an engineering program, and in turn, the students could be entitled to professional engineers by the Engineering Society of Denmark. However, in daily practice, there were still different ideas about whether they were real engineers or not. During the interviews, the topic was brought about in all the group discussions, and many of them mentioned their feelings of 'not being accepted as real engineers'. According to the students, they were not regarded as 'real engineers' by people from hard-core engineering, because the way they judge engineering is based on the technical contents in the learning subjects.

When asked how they understood engineering/engineers, most answers involved 'it depends'. For them, there were two perspectives to understand the concepts of engineering/engineers. From the traditional perspective, engineering was still associated with technology-related work with representatives of hard-core branches like electronics, construction, and mechanics. In turn, engineers were people who were working on hard-core engineering and had a general image of nerds. According to the students, this was also the image of engineers in the minds of average lay people in the society, which was the reason why they were not regarded as engineers. As discussed by the female students in group 5,

> S1: "Some people don't think that we are qualified to become engineers. They don't think that we have enough technical competencies. I think that we are, because there are so many different kinds of specializations within engineering, for example, building and mechanics, we should be one of them."
>
> S2: "Yes, I think that there are more things that have been put into the word engineering. Industrial designing is also one specialization within engineering. But people don't always know that there are different kinds of engineers."
>
> S3: "For many people, engineers are just engineers, people who are good at solving problems, and come out with a functional solution."
>
> (Group 5)

When asked whether they regarded themselves as engineers or not, students responded in very different ways to this yes/no question. However, their explanation to their yes/no answers shared similar ideas.

> S1: "Yes, I think that I am an engineer, but not the hard-core engineer. The title of engineer tells people what you are doing. It is quite difficult for many people to tell what industrial designers do. It can be easier for them to understand what you know and what you do if we say *we are engineers who are doing designing*."
>
> S2: "No, I won't regard myself an engineer. I would say that *I am an industrial designer with engineering knowledge*. In this way, I won't be associated with the nerd image of those technical engineers."
>
> S3: "I won't see myself as a pure engineer or a pure designer; *I am in-between*."
>
> S4: "Sometimes in life I do feel kind of hesitant when people ask me about my education. Shall I say that I am an engineer or industrial designer? I don't know what to say to them because I don't know the right title. So I just say that *I am an engineer in industrial designing*."
>
> S5: "The right title sounds very complicated, if I say that *I am an engineer in architecture and design with specialization in industrial design*. To elderly people, I just say that I am an engineer because it is difficult for them to understand what an industrial designer is. So it is easier to say that I am an engineer."
>
> (Group 6)

These seemingly different but factually similar answers showed a 'between' situation of these students. To the questions what they would like to be, most of the students replied 'both'. As quoted from one student,

> "We are going to be between architects and engineers, which means that we will have skills of both areas, and lots of things need to be changed when we establish a new area."
>
> (Nora)

From 'between' to 'both', an awareness of the need for making changes in both individuals and the profession could be witnessed. Most of the students believed that they were 'technically qualified' to be engineers, because engineering should not only be focused on technology. According to some of them, the ways of defining engineering and engineers should be updated and take diversity into consideration. And, their practice has brought about changes of values. As Lee said, 'now some of the engineering supervisors can see that even though what we are going through is more process-oriented, it is still engineering that we are doing.' In this way, some students regarded themselves as pioneers of making these different changes as well as bridging communication between two previously different professions. The learning experiences in this new branch of designing engineering changed themselves in perspective of understanding learning and career.

To conclude, through reflecting on both pros and cons of the learning environment, through the intervention between different values and practices that were brought by teaching staff from different professional areas, and through being a new type of engineers, students in A&D experienced a process of negotiation in the development of their professional identity.

8.6 Students' Perceptions on Gender Relations

From what has been reported in the previous sections in this chapter, the appreciation of the learning environment could be witnessed from the experiences of students of both genders after they went through the confusing periods, however, there were no impressive gendered patterns in the learning processes in A&D. During the research, female students were more interested in the gender aspects of my study,[13] because according to them, they had lots of discussions about gendered ways of learning and studying technology as females. Male students did not show special interest in this topic. During the interviews with the female groups, they often brought about the topic by themselves. However, during the interviews in one mixed group, when I brought about the questions of gender after the discussions of their experiences of learning, it did not lead to very much discussion.[14] In the following, I will report how the students at

[13] I introduced my project at the introductory day (the first day in the semester) with the presence of all the students.

[14] I sensed that students did not feel comfortable talking about gender issues with the presence of students of the other gender. Therefore I stopped questions relating to gender in the interviews with the mixed groups.

A&D in my research perceive gender roles, masculinity and femininity based on their life experiences.

8.6.1 Gender Roles and Identities

With the general equal participation of both male and female students, students could be themselves in the process of learning regarding speaking out individual opinions, communication with both teaching staff and peer students, and so on. In the break time, different informal topics were discussed, for example, female students talked about fashions and male students discussed sports and computers. As they said,

> "On the daily basis, it does not matter how we are dressed, we just wear whatever we want. Working here you can just be yourself. I don't think that people need spend energy thinking about that."
>
> (Group 5)

> "It is nice to have the girls around with different fashionable clothes; we see it as a good view. We can chat and laugh together; this is a nice and enjoyable atmosphere when working in the studio this way."
>
> (Group 2)

When talking about general femininity in the current Danish society, most of the female students referred to the topic of gender equality. According to them, there has been a visible observation of social changes in the past decades, which led to dramatic improvement regarding women's social status in both family life and job market.

> S1: "I think that the differences between men and women are getting smaller and smaller now. Many women can get top jobs, and some of them even have a family when they work on the top positions."
>
> S2: "I think that it will be more feminine if the woman can get top jobs."

> S3: "I think that if you get a top job like what a man unusually does, you really need to show that you are a woman."
>
> (Group 5)

However, gender stereotypes could still be witnessed when students talked about the opposite gender. Students of both genders mentioned hard-core technology and competition in career as men's tasks in the society. Female students used the terminology of 'soft' men when referring to male students in some study programs in humanity areas. Male students used the terminology of 'masculine women' to refer to female students in mechanical engineering.

Female students from groups 4, 5 and 6 talked about changes in the general gender roles in Denmark. Most of them grew up in families where their fathers had better-paid jobs and their mothers had traditional female jobs and did most of the house-caring work at home. Compared with their parental generation, they felt more equal in terms of women's situations. And they are prepared for having both career and family life for their future.

> "Things have been changed so much in the past decades. There are more and more possibilities for women to had a career. But still we need time and efforts on family. Maybe that's why we are preparing ourselves by organising and planning things from now on."
>
> (Jean)

However, they also discussed about the future work and the balance between family life and career pursuit.

> S1: "it is still difficult to see what it is. It is difficult to see how to make a family. We are female and we are the one who will give birth to babies, which takes lots of time."
>
> Interviewer: how about the plan of relating what you have learned to the work place?
>
> S2: "it is a difficult question. You are actually asking whether we will choose family or give priority to career. Because you can't have both in this kind of field of designing. If we are working on design, it is going to take lots of energy to focus

on the work because there is lot of competition in this field. It is hard to manage that with half time work."

S3: "it should be possible to work and at the same time have a family, but you should not have the highest ambition in work, because then it takes 50, maybe even 70 hours per week on work in order to reach this goal."

S4: "I don't think that any of us expect to become the best designer in Denmark, we would be very satisfied if we can have a job that we can do something based on our interest. That can also give us time to have a family."

(Group 4)

These women students chose to study technology based on their personal interest and ambition. They hoped to combine technology with something involving social concern so that they could contribute their intelligence to social development by helping others. However, the nature of the design work and the still-existing stereotypes in gender roles bring worries to them. When they saw difficulties in achieving balance between career and family life in the future, it difficult for them to build up striving plans in career pursuit, which also to some extent influenced their academic ambitions and confidence in the study life.

8.6.2 Possibilities for Changes

During the research, these female students in A&D also showed their thoughts about the existing gender bias and their wish for more changes. As one female student said,

"I think that some old values are still there when people say girls from some special education are more manlike. We should turn it the other way around, when we see more and more girls studying engineering, the education changes, it is not only for men any more..."

(Ann)

Awareness of benefiting the engineering profession by bringing in feminine values was shared by some of the female students in this study.

> "Talking about femininity, I think that it can be a good thing to have us in this kind of work. I know that it might sound stupid, but I think that we will make technology more attractive by bringing our feminine values."
>
> (Nancy)

Nancy and her group members (in group 4) had strong interest in studying technology as well as clear awareness of bringing feminine values into the design of technology. At the beginning of the semester, they expressed this expectation to their supervisor, as they said at the end of the meeting,

> "This is a really good meeting, we discussed a lot of interesting ideas, I think that we have a good start ... before we finish this meeting, I would like to say something more about our expectations, as we have discussed (in the group), we are 6 girls working together on this project, we are all interested in this topic.[15] We would like to say that we are very serious on this project and we want to be taken seriously as well. Please be straight and open to give us comments and critiques. Don't show sympathy because we are girls ... "
>
> (Nancy on behalf of group 4)

After the meeting, they told me that they actually never experienced bias from supervisors before. There were two reasons why they talked about it during this meeting, as they explained to me,

> S1: "this is because Mary[16] has heard some stories about that. She had a girl friend who studies in another engineering program. They said that sometimes professors will allow you pass no matter what, as long as you are a girl.

[15]This is based on the observation in the first supervision meeting in group 4 in the first week in February, 2004. At the moment, they were discussing different possibilities for the topic of their project. From this meeting, they got very positive comments on their plans of a topic from their supervisor. This topic was about designing a part of some medical equipment for a hospital. At the end of the meeting, Nancy said this on the behalf of the group).

[16]Mary was not present at that moment.

> S2: it is also because we just chose Industrial Design this semester, and we really want to take it seriously, so we think that it might be a good idea to set up this goal from the beginning…"
>
> (Group 4)

This expectation was respected by their supervisor during the course of the project. This group was in general satisfied with both of their supervisors regarding the way of supervision. However, sometimes their supervisors had difficulties understanding their feminine thoughts. In the first status seminar,[17] they presented the cartoons they designed as a way to giving instruction to the nurses in the hospital regarding how to use their product. This idea got some critiques from different supervisors.[18] When I asked them how they felt about it afterwards, they told me that they were criticized partly because this idea was not mature yet, and there were some technical problems. It was also because the method of using cartoons was not regarded as serious. It was easier for them to accept the former reason than the latter. After different discussions later on, they agreed to keep the cartoon after modification (a copy of the cartoon design that was used for the status seminar can be seen in Appendix 7).

During the interview, these female students also expressed their willingness to make changes in the engineering profession by bringing in feminine contributions, in spite of their lack of confidence in convincing others. As they said that they would like to make efforts on that so that they can prove it.

To summarize, both interview and observation data at A&D show the appreciation of student to the learning environment and to the general improvement gender equality in the society. However, gender stereotypes in the overall labour division still play a role in the ways people perceive the relations between gender and technology (for example, technology is still generally regarded as something men are better at than women by average people). This makes women's and men's experiences different in the process of studying technology. In my study, women in A&D, study, despite of the general equal situation as they said, had more concerns and worries in terms of how to study

[17] As described earlier in this chapter, in this seminar every group is supposed to make a presentation in plenum reporting their status of the project. They will get comments from their supervisors, other supervisors and lecturers, and students from other groups.

[18] All of teaching staff at present were male.

technology and bring their values to the development of technology than their male peers.

Summary

In this chapter I have reported the empirical study of my research at A&D. This description covers an introduction to the study programs from the perspective of educators, student's paths to this study and their experiences.

This group of students made their choice of this study with a driving force of their personal interest in learning technology, designing and developing creativity. As newcomers, in the first year these students confronted different new aspects in life: new learning subjects, new study methods, new ways of thinking and working, long work hours, stress from heavy workload and insecurity from getting used to a new environment. In this way, in their learning processes, they also experienced a transformation process from having difficulties in seeing the meanings of learning to appreciating learning in the PBL environment. The main reasons for this transformation were due to their gymnasium background, new subjects and new study environment. Their experiences in the later semesters witness an active learning process whereby students took the responsibility of their own learning and develop different strategies to improve learning opportunities for themselves. This is also a process of the active participation of students into the establishment of the learning culture.

The learning process in PBL is also a process whereby students develop a professional identity. However, the established image of engineering/engineers (hard-core branches) remained the barrier for the new engineering branch like A&D to be accepted as real engineering. In this way, students in A&D confront dilemma situation regarding clarifying their professional identity. Their experiences show that in the long run, engineering should broaden its definition and put itself into more categories rather than one.

The equal gender distribution of students in the study program makes students feel being themselves when studying technology. Nevertheless, gendered patters in the practices of learning can still be witnessed due to gender role in relation to technology: men are better at computers and technology and women are lacking experiences with computers in the childhood and adolescent years. This makes differences in women's and men's experiences in higher education. Women still have to make more efforts to bring their values in the work of technology and consider their future plan to work on technology. However, the data in A&D can be seen as indications of changes in engineering for the future, with more new inputs in terms of values and contributions from the participation of newcomers (some group of men and more women).

9

Gender and Learning in Engineering Education

In the last two chapters I have reported the empirical findings on the learning processes of engineering students in two departments in a PBL environment. In both science and everyday life, facts do not speak for themselves (Silverman, 2000). In this chapter, fieldwork findings are analyzed and interpreted by following three steps. In the first section, research findings are summarized, which is structured based on the presentation in Chapters 7 and 8. On this basis, research results are discussed in the second section by means of a comparative analysis in relation to the research questions, before the results are reflected to the theoretical frameworks for further discussions in the third section. This chapter concludes with a summary of answers to the research questions.

9.1 A Comparison

Before the research results are related to the theoretical framework for analysis and reflection, a brief summary of a comparison on the findings in the two research sites is presented in this section. These findings will be elaborated further in other sections of the chapter.

In Chapters 7 and 8 empirical data generated from two research sites — two different engineering departments — were analysed in the form of descriptive report. In the research process, data generation was guided by research questions, learning principles of PBL concepts, the theoretical framework, and some preliminary assumptions. With this foundation, I researched on reasons for students' choice of the engineering study and their experiences in the

Gender and Diversity in a Problem and Project Based Learning Environment, 265–293.

Research findings \ Research sites		EE	A&D
Path to Engineering		- Male: a natural path due to the closeness to technology - Female: a challenging choice based on interest in maths	Male and Female: Interest in technology, design and creativity
Being newcomers		Visible differences in the transition process: easier for male from technical gymnasium than male from math gymnasium and female	- Equal starting points for all students; - Hard transition process
Learning culture In PBL environment	Students' perception on learning	- Valuing understanding, meanings, practice - Appreciation of peer learning - Awareness of self-measurement	- Valuing meanings, reflection, and context - Appreciation of peer learning - Awareness of self-measurement
	Learning resources	- group work benefit the study of engineering, especially for female students, - learning from social activities - gendered forms in contributions to projects male: technical skills female: planning, management, organization	- group work benefit the study of engineering, especially for female students - confusion from different values from supervisors - gender differences in working style do not matter learning opportunities, but still exist
Professional identity development		- developing engineering identity studying EE in the workplace-imitated learning environment - learning engineering ways of thinking - certain patterns of interactions - different experiences for men and for women Male: close relation to engineering Female: being different, minority, adapting to the male-norm, dilemma situation in being an engineering and being a woman, becoming one of the guys	- developing professional identity studying in the PBL environment - dilemma situation when confronting different value resources from teaching staff with different backgrounds - developing a new engineering identity - bringing feminine values to the technological work
Students' perceptions on gender relations		For students of both genders: Masculinity : 'logical', 'decisive', 'analytical', 'structured' Femininity: 'gentle', 'sweet', 'elegant-mannered', 'intelligent'.	- women get more equal situation now More concern on this topic from female students: - worries about possibilities of keeping balances between work and family life

Fig. 9.1 Learning processes in PBL — research results in summary.

learning processes in the PBL environment with an intention to examine gender differences and patterns. Also examined were students' perceptions on gender relations based on their life experiences. Chapters 7 and 8 are structured based on the findings from the above mentioned aspects. A summery of the findings from the two research sites is outlined (see Figure 9.1), which is steered by the structure in the presentation of the findings in Chapters 7 and 8.

As Figure 9.1 shows, on the top rows are the information on students' choices of entering engineering and their experiences as newcomers. Information on both of the topics is mainly drawn from interview data. This information provides relevant knowledge on the backgrounds and the past learning experiences to these students.

The rows in the middle of the figure illustrate the learning processes of the engineering students in the PBL environment at EE and A&D, which are categorized into two parts: 1) learning cultures with respect to students' perceptions on learning and the multiple learning resources; 2) professional identity development. Information on these topics is mainly drawn from students' interpretation to their experiences in the learning processes as well as the interpretation from the researchers' point of view based on both interview data and observation data. The information is directly relevant to one of the main concerns of the research — engineering students' experiences in their learning processes in the PBL environment.

The information based on the listed topics in the figure is also structured with a gender perspective. In the top four rows, the information is organized in gender forms based on the observation and interpretation from the researcher's point of view. The information in the bottom row is based on students' perceptions on gender relations. The gender-based knowledge generated from empirical data is directly related to another aim of this study — influences of gender relations on the learning processes of engineering students in the PBL environment.

Therefore, this section provided a summary of the research results based on empirical findings described in Chapters 7 and 8. This review is also an attempt to provide a platform for the further discussion of the findings in the comparative analysis as well as the theoretical analysis in the following sections.

9.2 Life as Engineering Students

In this section, research results from the two research sites are discussed through a comparative analysis in relation to the research questions. This comparison is illustrated in Figure 9.2 and is elaborated in this section.

9.2.1 Learning Processes in the PBL Environment

In terms of learning processes, I examined 1) the learning cultures with respect to students' receptions on learning and the multiple learning resources; and 2) achievement of professional competences and identification through studying engineering in PBL environment. In the following, these aspects will be

<table>
<tr><th colspan="3">Research sites / Comparison of research findings
structures</th><th>EE</th><th>A&D</th></tr>
<tr><td rowspan="5">Learning processes In PBL environment</td><td rowspan="3">Learning cultures In PBL environment</td><td>Students' perception on learning</td><td colspan="2">Similar findings (few differences) – contextual knowledge, active learning, peer learning, seeking meanings, self-measurement</td></tr>
<tr><td rowspan="2">Learning resources</td><td colspan="2">Similar findings : multiple learning resources</td></tr>
<tr><td>- Resources for learning are mainly restricted to hard-core engineering
- conforming to certain ways of working
- few critical comments on PBL</td><td>- Resources for learning come from diverse social contexts, cross-disciplines collaboration is necessary
- diversity and creativity in working methods are valued
- critical and reflecting comments on PBL</td></tr>
<tr><td colspan="2" rowspan="2">Professional identity development</td><td colspan="2">Similar findings: identities development in the learning processes at two levers
- group identities
- professional identities</td></tr>
<tr><td>- Hard-core technology based professional identity = engineeirng identity
- 'Real' engineers</td><td>- Technology-based professional identity with creativity and social concern as contents
- not regarded as 'real' engineer</td></tr>
<tr><td rowspan="4">Influences of gender relations</td><td colspan="2" rowspan="2">Gender differences based on researcher's observation and interpretation</td><td colspan="2">Similar findings
- gendered features in doing project work : male are better at technology; female are better at planning, organization, management
- group work plays a supportive role for women to learn technology</td></tr>
<tr><td>- visible gender differences in paths to engineering, experiences of being newcomers, and different aspects of learning processes
- male norm: more privileges for men and more barriers for women
- technical skills based norm: women' values are not appreciated yet
- men: closeness to technology women: become 'one of the guys'</td><td>- few gender-based differences in paths to engineering and experiences of being newcomers
- gendered features in different aspects of learning processes do not play negative role to neither men nor women
- contributions of both genders are appreciated and valued
- a diversity of femininities and masculinities</td></tr>
<tr><td colspan="2" rowspan="2">Gender relations from the perspectives of students</td><td colspan="2">Similar findings:
- recognition of overall social changes in terms of gender relations
- women in both cases showed more concern in women's rights of pursuing career and possibilities of keeping balance of work and family life</td></tr>
<tr><td>Regarding gender identities, students of both genders refer to ways of dressing, manners, and gendered professions</td><td>Little gender constraints in the perception on femininity and masculinity</td></tr>
</table>

Fig. 9.2 Comparative analysis of the findings.

discussed respectively through a comparison between findings from the two research sites.

(1) In general, this study witnessed similar pictures in the two research sites regarding their strong awareness of reflecting on learning based on their experiences as well as their perceptions on learning.

Their reflections on learning can be summarized in the following aspects.

— Learning is more than the reception of factual knowledge or information; rather, it involves more aspects such as goals, expectations, understanding, application, meanings, insights, values and so forth.

— The process of seeking understanding, meanings and values is closely related to practices and activities in a context, which involves participation, communication, interpretation and reflection.

— Peer learning through collaboration in team work is considered as an efficient way of learning technology and preparing themselves for the workplace culture.

— Self-measurement in the learning process is regarded as an important aspect to seek the meanings of learning. Grades are not regarded as the only objective and way of assessing learning. Instead, a focus on measuring learning as a process was witnessed from the points of view of individual learners.

For students from both EE and A&D, studying engineering in the PBL environment, it is rather essential to take consideration of methods and strategies of gaining knowledge, the application and context of knowledge, and the meanings of the knowledge.

(2) Based on the learning principles of PBL concepts and different learning theories, this study made an assumption that the learning resources for studying engineering in the environment of PBL, Aalborg Model is mainly derived from three areas: process of problem-solving and doing project in groups, attending lectures, and getting guidance and assistance from supervision. As has been reported in Chapters 7 and 8, the empirical work from both of the research sites has shown that these three aspects played constructive roles as the major learning resources for students to study engineering. In addition, the data from both of the engineering programs illustrates more resources than these three aspects, as Figure 9.3 shows.

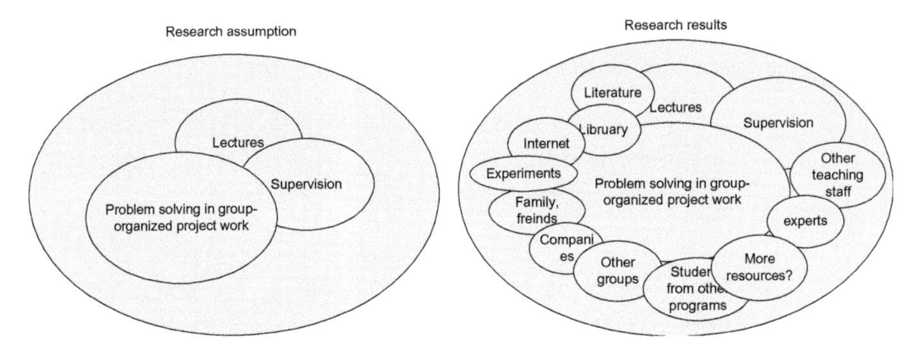

Fig. 9.3 Multiple learning resources in the learning processes in the PBL.

For engineering students in the PBL environment, the core of the learning process is to do projects in groups with the aim of solving problems. In my study, interview and observation data from both EE and A&D mirror to each other with regards to the learning process of getting through project work in groups (without considering gender differences, which will be discussed in 9.1.2).

To get the project started, students need to search for the information on the background and context, to find relevant literature, to read theoretical articles, to discuss with supervisors or people who know the area, and they might also need to contact industries or companies for interviews or observations to gain field knowledge. When they have collected enough material, they start to analyze the situation and formulate the problem. The next stage is to find out how to solve the problem and choose one of the solutions, and this involves same procedures of searching, reading, discussing and writing. In this process, they are facilitated with the knowledge from the literature, lectures, and supervision; however, they are expected to relate these different knowledge resources to their project. They need to develop different strategies to gain theoretical knowledge, methods, and context knowledge in order to solve the problem. This is also an interactive process both within the group and with the real world. For example, group work involves discussing, reaching agreements, writing, etc. which demands the awareness and skills of communication, collaboration and management. In addition, they share experiences and resources with other project groups, they turn to other supervisors or experts when they need more information than what can be provided by their supervisors, they use personal network (family, or friends) to build up contacts for research, and so on. Project works normally involve across discipline knowledge, which in turn, involve searching resources of information from different areas.

Therefore, research findings in both research sites illustrate that when studying engineering in the PBL environment students take active roles in developing learning strategies and take responsibilities of managing their own learning. This is also a way that students manage their own learning with self-awareness of learning goals and expectations.

Some slight differences were observed regarding learning resources in the two research sites. At EE, in spite of the multiple resources, the sources of learning are relatively restricted to the hard-core engineering areas. In addition, the nature of the study defines some certain ways of working in doing the project for students to follow. During the research process, there were few critical comments from students at EE towards the PBL environment. Comparatively, at A&D, collaboration across-disciplines are sometimes necessary and encouraged, which involves learning resources from diverse social contexts. The unique and diverse working methods as well as creativity are valuable in the project work. Based on their experiences in the learning processes, students brought about different critical thoughts and reflections on the PBL environment (see Chapter 8.5.1).

(3) The research in both of the research sites shows that studying engineering in PBL involves a process of equipping students with professional competences in a certain learning culture, which can be seen as a way to develop identities.

In both EE and A&D, through doing every project, students develop a group identity together in the process of establishing daily practices, routines, patterns of communication, ways of dividing tasks, handling conflicts and reaching agreement, and norms. This can be exemplified from the way they built up either written or hidden collaboration agreement, develop strategies to make best of supervision and so on. In general, ways of developing group identity differ from one project to another, and from one semester to another.

Visible differences had been witnessed at EE and A&D regarding the development of the professional identity. At EE, students are identified (by lay people or non-engineers) and they identity themselves as ‘real engineers’ based on the hard-core technological skills and problem-solving oriented professional responsibilities. For them, they are the 'real engineers' due to their close relation to technology in practice, and to solve real problems can be regarded as a way to identify their profession.

In addition to the technical skills to some extent, students at A&D have concerns with creativity and innovation, the social application and impact of technology, communication and dialogues with the society to achieve feedbacks in their study. Therefore they develop different professional skills from students at EE. However, they are not regarded as 'real engineers' either by outsiders or by insiders of the engineering community. Instead, they are seen as 'soft' engineers by students at EE to distinguish the new engineering programs between from traditional hard-core engineering programs at the university.

9.2.2 Gendered Experiences in Learning

This study examined the influences of gender relations on the learning processes of both male and female engineering students in the PBL environment. The following discussion is based on the knowledge generated from two angles: the observation from the researcher's point of view in a gender perspective, and the perceptions of the students on gender relations.

From a gender perspective, different pictures at EE and A&D were observed regarding women's and men's experiences in the aspects of study life, from the paths to engineering, social academic backgrounds, being a newcomer, experiences and strategies in the learning processes, and identity work process. These aspects construct different gendered learning cultures in the two research sites.

The research at EE identified a 'natural' process for male students to enter engineering due to their tinkering experiences, interests in technology and the suitable gender role for taking engineering as the future occupation. For the limited female students, their paths to engineering were either influenced by their fathers, boyfriends or other social relationships that were engineers, or motivated by the strong interest in math and good performance in science subjects at upper-secondary schools. The experiences of being newcomers in the engineering program witnessed a harder transition process for female students than for their male peers due to their lack of technical knowledge and unfamiliarity with the study form.

Studying in PBL environment, although developing social skills are in general regarded as useful learning strategies by students of both genders, technical skills are all in all prioritized as engineering competences in the minds of male students and possibly most male teaching staff[1] at EE study program. In the group work, these female students are in general better at managing, planning, organizing, and coordinating communication. However, these features can not replace their lack of technical skills. Female students need to make special efforts to catch up and prove their academic performances. The adaptation process is also a process in which they learn to do things in an engineering way, which is also a male engineering way. When they are eventually accepted by their male peers and male teaching staff, they have become 'one of the guys'.

Gender differences can also be witnessed in developing the sense of engineering responsibilities and working styles. Some features, such as solving problems and doing projects in the form of teamwork, are manifestly recognized as strategies of building up professional confidence and belief for career pursuit by the male informants; whereas, for female students, it is mainly education they are trying to get through based on their interest in science subjects.

At A&D, few gender-based differences were observed regarding their paths to engineering and learning experiences in the first years. Students of both genders followed similar trajectories to this study program based on their interest in technology, design and creativity. They shared equal starting lines and experienced similar paths from being newcomers to more experienced in managing learning, and developing professional identities in the learning process. A diversity of femininities and masculinities are witnessed in the daily practice of learning.

Nevertheless, gendered features in doing the project work can still be identified at A&D, for example, male students are in general regarded better at technology due to their early access to computers, Women's features of working in groups are similar with those at EE. In general, women are better at planning, keeping schedules, being systematic, and managing. Compared with research results from EE, there are two main differences concerning the influence of gendered features on the learning process: 1) at A&D, the gendered features in different aspects of learning processes do not play negative

[1] This message is obtained from indirectly observation from the students' experiences.

roles in the learning process of to either men or women; 2) at A&D, values and contributions of both genders are generally appreciated in the learning culture.

With respect to students' perceptions on gender relations, both similarities and differences can be identified from the two research sites. Overall social changes in gender relations have been recognized by all the students concerning gendered labour division in the family life and an increase of women's rights as well as the awareness in the pursuit of career. The topic of women's equal rights was specifically discussed by female students from both the cases. However, when the career is in relation to technology (especially hard-core engineering), it brings conflict with femininity. Accordingly, compared with their male peers, female students in both EE and A&D have more concerns and worries about possibilities of keeping balances between work (especially engineering work) and family life in the future. In their reflection on gender identities, students of both genders at EE referred to ways of dressing, manners and professions, whereas few gender-based constraints were observed from students at A&D regarding their perceptions on overall masculinity and femininity.

Therefore, gender relations play a role on the learning process of engineering students in PBL environment. At EE, gender relations make the learning processes different for men and for women by bringing privileges to men and obstacles to women. Gender differences might not be visible on the surface when female students make compromises by adapting to the male norm in the learning culture. At A&D, the gender differences do not seem visible, although they can be observed concerning students' ways of working and communication. In general, gender relations do not matter negatively since working on technology at A&D does not conflict with being a man or woman. Awareness of gender differences can even bring benefits to the project work when values from each side are appreciated and made use of. Especially, this positive function can be exemplified by female students' efforts of bringing feminine values into the design of technology.

To conclude, in this section research findings from the two research sites are discussed through comparative analysis, which is structured in relation to research questions. This structure provides a preliminary response to the research questions, which also paves a stage for the theoretical analysis in the following section.

9.3 Gender and Learning Analysis

In previous sections, the findings were summarized and analyzed based on the report of empirical work in Chapters 7 and 8 and a comparison of the findings between the two cases. In this section, the empirical work findings will be discussed and analyzed in relation to the theoretical frameworks that are developed in Chapters 3, 4 and 5 as well as the learning principles of PBL concepts that are discussed in Chapter 2. The discussions will lead a way to the answers to the research questions.

9.3.1 Learning as an Engineering Student

In Chapters 2 and 3, I reviewed the learning principles of PBL environment in engineering education and different academic scholars' works on theories of learning in constructivist-sociocultural approach. In the following contexts, the research findings will be analyzed in relation to the relevant learning theories and the learning principles of the PBL environment.

Examining the learning process from three levels

The perceptions of understanding learning directly influence the ways of assessing learning (Rogoff, 1997). Accordingly, a model of examining learning at three levels was developed (in Chapter 3) with an aim to achieve a comprehensive understand of learning. This model encompasses levels of individuals, communities and institutions, and broader social cultural contexts as well as the interaction among these different levels. In the following, three levels of the model of understanding learning will be discussed in relation to empirical findings and the intertwined relationships between the three levels are emphasized along the way.

Individual/personal level

For learning at the individual level, deep understanding is placed as one of the main focuses of learning in higher education from the point of view of critical scholars. As Barnett (1994:102) defines, deep understanding exists when the learner has a particularly clear perception. Based on the research findings in this study, this emphasis is corroborated with both the design of PBL concepts as the learning environment and the practice in students' learning processes.

In addition to individuals' acquisition of knowledge, learning also involves participation in activities as well as the reflection on the experiences. In the review of experiential learning theories, Brookfield (1983) summarizes two different ways of using of the term 'experiential learning'. 1) From the perspective of facilitating learning, and through the curriculum design, students are given chances to acquire and apply knowledge, skills and feelings in an immediate and relevant setting. 2) From the perspective of individuals, experiential learning occurs as a direct participation in the events of life. Instead of being provided by formal educational institution, learning is achieved through reflection upon everyday experience.

From the viewpoint of my study, both of the ways are reflected in the practice of PBL Aalborg Model. The learning principles of the PBL environment at AAU illustrate the intention of providing contextual knowledge and meaningful learning. At the beginning of the first year, engineering students were provided a course on Collaboration, Learning, Project Management (CLP), where they were taught learning methods and some basic learning theories such as Kolb's (1984) experiential learning circle. This course, according to most of the students in my research, helped develop self-awareness of learning and develop learning strategies to handle different situations when doing projects in groups.

From the individual perspective, having sufficient chances to participate into activities and practice, students in my research showed strong awareness of learning from experiences. When students actively enter an experience, they encounter chances to understand themselves and the context as well as contents they are engaged with. This learning process involves loops of experiences and reflection at both cognitive and affective levels. This can be witnessed from both the general increase of self-awareness in learning and the reflection in project work as well as in daily practice.

At both EE and A&D, research findings identified the application of this learning circle to the daily practice. For example, in the lecture aspect, the assignment from each lecture provides students chances to test the ideas and theories through contextualized practice, from which students encounter concrete experiences. Discussions about these experiences of practice-based assignment with peer students and lecturers provide feedbacks and reflections that lead to deep understanding.

The semester-long project work is also closely associated with experiential learning. At the beginning of each project, students in the new groups would reflect upon their good and bad experiences from the past projects and group work. They analyze the strengths and weaknesses at both individual and group level. Based on the reflection on experiences and their shared interest and learning goals, they moved to conceptualize about group theories by establishing new practices and ways of working, which will be tested in the new experiences. The discussions with group members and supervisors provide more experiences and chances of reflection for students to know their behaviour, feelings and thinking as well as others. The feedbacks and comments motivate further reflection and generate more theories for guiding practice in group work.

Community/institutional level

As discussed in Chapter 4, from a sociocultural perspective, theorists like Jarvis (1992), Lave and Wenger (1991), Rogoff (1995), Wenger (1998) perceive learning as constant and meaningful engagement in shared practices, in which they interact with others and with the context. These practices are thus the property of a kind of community created over time by the sustained pursuit of a shared enterprise. In particular, Wenger (1998) writes about three dimensions to cultivate a community of practice for learning (the domain, the community and the practice, see Chapter 3). Relating to engineering education, there are different levels of communities of practice, for example, university, study programs, and project groups. Specifically, my study examines how learning is achieved through cultivating communities of practice in the daily practice of doing project work in groups at two different engineering programs.

Learning through participating into different communities of practice is indicated by informants of both genders as a beneficial way of studying engineering. Particularly, a supportive community of practice is referred to as an encouragement to keep some women from dropping out. Students play an active role in managing their own learning: they learn to take initiatives, set up learning expectations, formulate learning goals, seek various resources for learning, develop learning strategies and evaluate learning outcomes. This exemplifies the self-directed learning that is advocated by different educators as a meaningful way of learning (Jarvis, 1992; Zimmerman and Lebeau, 2000).

This learning process also covers the items that are emphasized in the assessment of learning in the sociocultural approach: individual roles, changing participation, coordination with others, the nature of the activity and its meaning to the community (Rogoff, 1997). Relating this process to the six indicators of evaluating learning as suggested by Rogoff (1997) (see Chapter 3). I see this study as an example of evaluating learning and development from a viewpoint of transformation of participation. Studying engineering in the PBL environment, at first, students as newcomers need supports for all phases of doing project in groups, especially how to formulate and analyze research questions and how to manage the project process. Gradually, they begin to take on responsibilities for the project work, like taking turns to be group leaders, participation into dividing and taking fulfilling different parts of work. The collaboration in the group work also helps to promote the sense of responsibilities, to develop the readiness for commitment to the communities' practice and to make different contributions to ongoing activities as they transform their participations. They gain deeper understandings of the project work and play more active roles in designing and planning of the projects. Project work proceeds with their growing interest, active involvement, and flexible and positive attitudes to make adjustment in the carrying-out process. They bring in different values into the shared practices from the reflection on their past experiences in different contexts. I witness this as a fruitful learning process whereby the learners develop different competences and make changes in themselves as learners and in the communities' practices.

Sociocultural level

As has been discussed in Chapter 3, in addition to the mastery of technical skills, involvement into activities, reflection on one's experiences, development of communication and collaboration skills, learning also involves becoming a person in different communities as well as in a broader society. From the perspective of Bildung, this becoming involves both command of knowledge and learning possibilities, as well as a process of self-formation and self-cultivation (Bauer, 2003). Therefore, educational institutions carry responsibilities of providing learners with knowledge, abilities of reflection and cooperation as well as a self-development process to become a responsible citizen in the society (Henriksen, 2006).

Previous research in traditional learning environment identified visible distance between being an engineering student and getting to the engineering workplace (Du, 2006). Based on what has been discussed on the research finding, I can reach a conclusion that a student-centered learning environment can prepare students with more chances to gain not only scientific knowledge, technical skills, but also capabilities of managing project and team work as well as professional responsibilities in order to prepare themselves for the workplace. It also provides a milieu where learners can interpret their experiences in the learning processes as meaningful. If Bildung is set up as a goal for education in the learning society (as argued in Chapter 3), the PBL environment as an example of student-centered learning environment at the engineering university fulfills the goal of education.

Meanings seeking and identity development in the learning process

As has been discussed in chapter 3, two major characteristics of the contemporary constructivist-sociocultural perspective of examining learning are (1) the highlight of seeking meanings as one main objective of learning; (2) the recognition of the significance of identity development in the learning process. Relating these focuses to the Bildung concepts and engineering education, engineering students are expected to learn what is engineering in order to become an engineer (Henriksen, 2006). Therefore, in addition to the mastery of technical knowledge and skills, studying engineering at the university involves engaged experiences with the professional culture as well as knowledge about what it means to become an engineer (Du, 2006).

As what has been summarized from the research findings, studying engineering in the PBL environment, students draw upon multiple learning resources. Students take active role creating learning opportunities and managing their learning processes. A process of seeking meanings through learning is indicated, which is appreciated by students.

Findings from both EE and A&D show that at different levels of communities of practice in engineering study, there are attitudes, principles, responsibilities, values and expectations that provide messages in perspective of constructing an identity. My research have observed that in the PBL environment, the learning culture at different levels of communities of practice (from the study programs to different project groups) encompasses academic training for expected competencies, social interaction routines and established practices for doing professional work. Therefore, this study identifies that studying engineering at the PBL environment involves not only mastering technological competencies, but also an identity work process, whereby students develop a sense of belonging to different levels of communities and achieve a certain membership.

Based on different tasks and participants as members, ways of cultivating communities of practice change from project to project, from group to group. Accordingly, the ways of achieving memberships and identity at the project group level of communities of practice are diverse. Students have the awareness of the fact that the group-organized project work challenges the development of ones' personal identity and that they have to work on the development of this identity in a group context. The experiences of students also show the need for flexibility, change and balance in order to make each community of practice (project work level in this case) function well so that everybody can learn.

Relating research results to the theory of communities of practice (Wenger, 1998), both of the cases can be seen as examples of participating in and achieving valid memberships in communities of practice at engineering educational institutions. Findings at EE identify a process of participating in the practices of engineering communities, where students are trained engineering competences and learn to become an engineer. Findings at A&D also witness a process of developing a professional identity through participating in the communities of practice. However, they are confronting the dilemma in the identification of professional identity. At a formality level, as a result of negotiation, A&D has been accepted as an engineering program and students can have memberships of the Danish Engineer Society. Nevertheless, at the cultural level, they are not regarded as 'real engineers'.

From the perspective of communities of practice, reasons for different situations regarding the development of professional identity in the learning processes of engineering study at EE and A&D can be analyzed as the following. 1) The values in engineering community at large are based on the established practices in 'hard-core' engineering, which is closely linked with technology. 2) Accordingly, as the 'old-timer' of engineering community, 'hardcore' engineers are set up as the public ideology of engineers in general. According to the majority of the informants in my research, the public ideology of engineers in the current Danish society is symbolized by vocabularies like 'working with computers', 'problem-solving', 'logical', 'structured', 'focused' 'rational', 'analytical', 'disciplined', 'nerdy'. Hence, 3) in the engineering educational institutions, students are trained to obtain the expected engineering competences so that they think and work in 'the' engineering way, which is based on the norm of 'hard-core' engineering. 4) Different from EE, the professional competences that students at A&D achieve are not only based on hard-core technology, but also related to people and society. Differences in the domains of the two communities (institutional level of the engineering community) lead to different practices and different community identities. By developing a different kind of engineering identity from that at EE, they are not regarded as 'real' engineers.

In this way, the two cases of my research provide examples of two different ways of participating into communities of practice. At EE, new members' participation is characterized by mainly following the established practices, fitting into 'the' engineering way of thinking and working, and reproducing the values and norms in the 'hardcore' engineering culture. In the historical development of engineering education, the community culture is established by the old-timers like the teaching staff, former students as well as influenced by the workplace culture. Therefore, students at EE can be regarded as the 'conformers' of engineering community.

In a new community of practice like A&D, all the members participate in the establishment of shared practices, principles, responsibilities, and values. This exemplifies that learning for individuals is more than conforming to the established practices and values, but also involves active participation in the cultivation of new practices. In this way, engineering students at A&D can be regarded as 'challengers' in that they actively contribute to the establishment of a new community of A&D and they bring new values into the overall engineering competences and engineering culture.

These two cases — 'conformers' vs. 'challengers' — also provide evidences in relation to the discussion of power issues in communities of practice, which is a topic that remains vague in Wenger's work (1998) (as argued in Chapter 5). In general, perceiving power from the productive perspective (as discussed in Chapter 5) (Masschelein and Ricken, 2003), the learning process of participation (no matter as 'conformers' or 'challengers') in communities of practice, and the achievement of professional identity can be interpreted as empowerment for both individuals and communities, in that it promotes the development of both. However, these two cases also illustrate different pictures regarding power relations between individuals and communities. From the production point of view of understanding power (Masschelein and Ricken, 2003), being 'challengers' (students A&D) implies more powerful situation than being 'conformers' (students at EE) in the individual/structure relations.

PBL concepts as an educational model in theory and in practice

One of the main aims of this research is to examine the practices of PBL concepts as an educational model from the perspective of students' experiences in the learning processes. The relationship between the learning principles of PBL Aalborg Model, learning theories and empirical findings is shown in Figure 9.4. A learning model (in Chapter 3) is established based on relevant theories that have been related to the learning principles of PBL Aalborg Model (elaborated in Chapter 2). Both the model and the learning principles are used as references in the analysis of empirical data.

As has been discussed, the research findings in this study witness a match between the theoretical design of PBL environment and the learning experiences of engineering students in practice. Students' perception of learning and their active learning process fit into the learning principles of the PBL concept and expectations of the educational designers at Aalborg University.

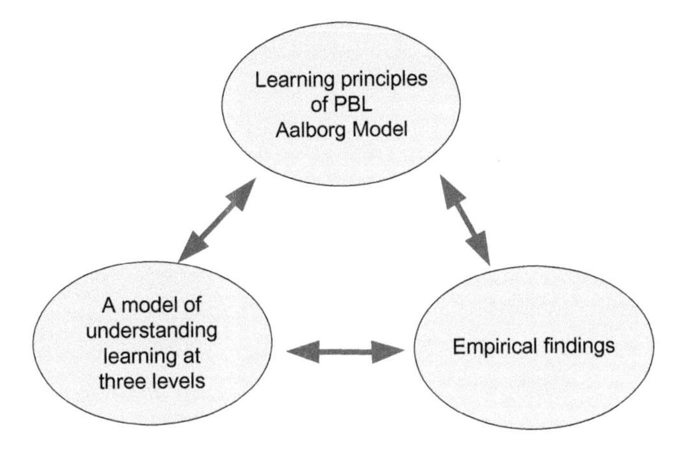

Fig. 9.4 Relations between learning principles of PBL, Aalborg Model, the model of understanding learning, and empirical findings.

A reciprocal relation between the empirical work and the learning model is signified as well. The learning model developed in this study provides an analytical tool to understand and interpret the empirical data. It also provides indicators to access learning from a constructivist-sociocultural learning perspective. The findings indicating the engineering students' experiences in their learning processes in PBL provide examples for the application of the learning model regarding understanding learning from three levels.

9.3.2 Gender Analysis

To examine learning from a gender perspective is one important aim of this research. In Chapter 5, an analytical framework was established by relating learning theories from constructivist-sociocultural approach (as discussed in Chapter 3), gender theories (as discussed in Chapter 4) and feminist works in engineering and technology areas (as discussed in Chapter 5). In this section, this framework will be discussed in relation to the empirical data, which is also a way to answer the second research question in this study. Responding to the analytical framework, this analysis is structured into three levels: individual, institutional and sociocultural level. However, different from the orders of presenting the three levels in Chapter 5, the following discussion will start with analysis at the institutional level, in that 1) it is one of the aims of this study to examine the individual-community relation concerning learning and gender; 2) organizational culture in communities of practice has been specially addressed by different feminist scholars (Harding, 1986, 1996; Gherardi, 1995; Kvande, 1999; Heyes, 2000) as an important factor influencing men's and women's learning differently; 3) the review of the findings shows that the organizational culture plays a role in bridging the understanding of individual experiences and social cultures; 4) the identified gendered patterns in both organizational culture and the individual experiences can be traced back and will lead to a conclusion at the overall sociocultural level.

Institutional level — gendered communities of practice

The institutional level in the analytical framework is about the gendered structure in organizations and gender roles within different contexts. As discussed in Chapter 5, it has been argued by different feminist scholars that gender contracts in the society as a whole define technology as a male task and an engineering university as a male location. Accordingly, engineering universities (in particular, engineering programs within the university) can be seen as both the moderate scale of community of practice and the institutional level of gender contracts. Feminist scholars propose two major factors — gendered proportion and norms in the organization — regarding the construction of the organizational culture from a gender perspective. Both of the factors can be related to the interpretation of the different pictures that have been observed in the research sites of EE and A&D concerning gendered experiences in learning.

According to Heyes (2000), the proportion of women students and teachers in an educational program remains one of the most vital factors deciding the culture. In the description of women's participation in historically-male enclaves, Kanter (1993) writes about women's experiences of being isolated, visible, and gender stereotyped when they engaged in work sites where they were the numeric minority, in particular, in gender-inappropriate occupations (Yoder, 1991). Relating to my study, the program of EE is witnessed to be a male-dominated environment and the token situation applies with women's participation lower than 5% (since the terminology of 'tokenism' is used by Kanter (1993) to refer to the situation of being a member of a subgroup with fewer than 15% of the whole group). Being a minority group, they are visible, observed and evaluated. As Kanter describes, they are not evaluated like one of the dominants (the terminology Kanter uses for the male majority), but as a different group from the male domain as a contrast. In my research process, all of these female students at EE expressed their wish to have some female students around either for mental support to gain professional confidence, or for a feminine social atmosphere. A new aspect that is found in my study is that this visibility and ways of evaluation of women's presence will last until the time when the tokens manage the adaptation process and 'disappear' in the majority group as 'one of the guys'.

At A&D, with more women's participation, a friendly learning atmosphere is established where students of both genders can be themselves at the same time working on technology. This reflects Heyes' (2000) notion that gender plays an influential role in shaping women's opportunity for learning in an institutional context. It also exemplifies how the proportion of female students itself can contribute to women's feelings of belonging to the community in the learning process and to their overall comfort with the learning environment.

Therefore, number (gender proportion, to be explicit) as an important factor influencing the establishment of learning culture, matters in the way of building up a friendly atmosphere in which learners of both genders can feel comfortable. However, by this statement, I do not mean that the high percentage of female participants in a community automatically guarantee a supportive culture for women's learning, because as Harding (1996) puts, in some formal educational settings, special values are placed on certain kinds of knowledge and ways of knowing. I agree with what Salminen-Karlsson (1999) states that the numerical composition of women and men does not decide the basic gender values of an occupation, in that the two genders have different amount of power. As concluded from her research findings in an educational reform with an aim of recruiting women, changing the numerical gender balance in a male-dominated area do not necessarily threaten a masculine dominance on the level of gender symbolism.

In addition to the gender proportion, cultural values in an organization remain another essential factor defining a gendered culture. Take an example of EE engineering, being a historically male task in many western countries, the culture of which defines the hard-core engineering education as a male dominated sphere. Findings of my research did not witness different pictures regarding the male centeredness in the PBL environment from what has been identified in a traditional learning environment as depicted in previous feminist works (Hacker, 1989; McIlwee and Robinson, 1992; Tonso, 1996; Berner and Mellstrom, 1997; Salminen-Karlsson, 1999).

The research findings at EE identified an association of the cultural identity of EE engineering with masculinity in Denmark. By choosing a traditional male profession, female engineering students at EE become 'different' or 'special' in the group of both women and men. The masculine image of EE engineering identity also shapes men's and women's learning experiences differently in the community of practice. The gendered experiences as newcomers identify some hidden expectations in terms of technical skills, which are based on the past experiences of the majority of male students. This brings about difficulties and inconfidence to female students because they did not have the same starting point as their male peers.

Different contributions in terms of female and male features can be witnessed in the process of doing project work. The established ways of working and communication in the engineering culture play a role in engendering the patterns of participating in the activities and interpersonal interactions. Implicit values are based on male experiences and interests (hard-core technology). The specific skills of women are not yet identified as engineering competencies in engineering practices. These culturally defined expectations, competencies and values on the knowledge in the engineering community make women's contributions less appreciated and therefore invisible. This engenders the culture of hard-core engineering knowledge as masculine and allows little room for the practical and societal implications of technology as well as other social concerns, which are associated with feminine values in western cultures (Harding, 1996).

The case of EE can be seen as an example of how the historically constructed male norm in engineering still defines the culture of EE engineering in the student-centred learning environment in the current society. The case of A&D provides an example of different gendered culture. With few established practices and values, a new community culture is being constructed with the participation of all the members when they bring new values based on their past experiences, their own characteristics in working, their contributions, and so on.

Therefore, concerning the gendered culture in the communities of practices, what can be concluded from the comparison of findings from the two research sites is (1) gender proportion plays a direct role in defining the learning culture; (2) norm and values decides the distribution of power to different social groups within communities of practice; (3) the nature of the study matters in that it defines some established ways of working.

Individual level — to be or not to be

In this study, the analytical framework at individual level is focused on the identity formation through interaction in the social context. These interactions are affected by both the organizational culture and the prevailing conception of gender. According to Wenger (2004), when participating into different communities of practice at the same time and obtaining multimemberships, learners might confront conflicts. Based on this, I had an assumption that female engineering students might confront dilemma situations between being an engineer and being a woman. This can be confirmed by the empirical findings when this study witnessed different pictures at EE and A&D with respect to gendered forms of identity construction and management in the learning process. In addition to this, this study finds that except for male students at EE, all the other students (including female students at EE and students of both genders at A&D) confronted dilemma situation in the identity management process.

The empirical findings at EE illustrated different situations for male and for female students in their identity development process. It is natural for men to study hard-core engineering and become engineers due to their appropriate gender role. For female students, they become boundary-breakers by entering a male sphere in order to pursue their academic ambition. Being tokens in a male-centred environment, they need to adjust to the engineering ways of working and to become 'one of the guys'. This finding is in tune with what has been reported by previous feminist work (Gherardi, 1995; Kvande, 1999; Salminen-Karlsson, 1999): the organizational context and the environment of communities of practice at workplace are created differently for women from those for men, which puts female engineers into a dilemma situation in following female gender identity and achieving professional engineering identity.

From being different to being accepted, this study witnesses a general adaptation process of these female students. This process cost them special efforts to manage the two conflicting identities, which is something their male peers need not do. However, these women's extra identity work and efforts as the cost of studying engineering has not been recognized as a problem by neither these women themselves nor their male peers and teaching staff. This was due to the taken-for-granted practices and established values in the engineering community. With courage and bravery, they took challenges in life by breaking the boundaries and entering a non-traditional profession. Their strong wills and capabilities are evidenced through their efforts to handle dilemma situations and to go through the hard-core engineering education.

From a gender perspective, this study perceives the construction of masculinity and femininity as a negotiation of gendered meanings through interaction in a social context. In this way, EE can be seen as an example, from which my study witnessed a strong tendency of negotiating femininity in order to keep a balance between femininity and being an engineering student. Studying engineering in an environment where the practices and values are based on male-norm, these female students developed different strategies to create chances for themselves to learn. Some women chose to 'put the discussion on the table' from the beginning of the project, by speaking out their own strengths and weakness as well as the expectation regarding learning. Some chose to invite another female student to work in the same group. In general, when they reached the later semesters, they became more experienced in handling different situations as being minority; they managed to cooperate well with male peer students and developed good friendships with them. They proved that they were qualified in terms of intelligence, technical competence, and mental strength.

At A&D, both male and female students take active participation into the community of practice. Being a member of this community is being oneself — they do not need to think whether they are a man or woman when they work together. With the increase of women's visibility and active participation, a new engineering culture is being established — being a design engineer is not conflicting with being a woman. A diversity of femininity and masculinity co-exist in the construction of the community culture. However, students of both genders at A&D are experiencing the dilemma situation in the identity development — to be engineers or not? They do not see themselves as a traditional 'engineer' because they want to identify themselves in a new way rather than fitting into the traditional ideology of engineers. As a compromise, some students identify themselves an industrial designing engineer.

In the research site of A&D, this research observed a negotiation process of bringing new values into the engineering community. The disagreement among teaching staff, who have different professional academic backgrounds regarding what kinds of skills and competences they are expected to achieve, made it difficult for students to identify themselves. Confronting this dilemma, students developed different strategies to handle the disagreement so as to achieve their own learning goals. Some group have the awareness of setting up their own learning goals and the aims of the projects. Some groups learned to 'figure out what is the most important and useful parts to the projects and say no to some of the suggestions by supervisors'. When facing serious disagreement from two supervisors, some groups chose to arrange meetings with the presence of both sides for discussions in order to reach consensus at certain levels.

In conclusion, the conflicts arising in managing different identities in the learning process to a great extent reflect how the norm, values and stereotypes in the local communities of practice shape individual experiences. At EE, it reflects the close link between male gender role and engineering identity and the gender stereotype in people's mind. Findings at A&D reflect the strong association of engineering with hard-core technology and the stereotype of professions. Male students at EE do not confront conflicts in managing professional identity and gender identity because they are working on hard-core technology, which is placed as the core of engineering; and because they have the appropriate gender role, which is set up as the norm of engineering identity. Different from them, female students at EE, and students at A&D are regarded as 'the others' in engineering community.

Cultural level — what count as engineering

The sociocultural level of the analytical framework refers to the general conceptions of gender prevailing in society. Following Harding (1986) and Hirdman (1990), they are conceptions of women's and men's inherent characteristics and their proper places, tasks and behavioral pattern. The assumption is that gender roles have impacts on the establishment of shared practices and values in communities of practice (especially engineering community) and on the development of individual identity and the principles of social behavior.

Based on the discussions at institutional and individual levels of the analytical framework, I have summarized that the dilemma situation in the identity management process will result in different conflicts in managing multi-memberships. I argue that the conflicts in the two cases both derive from the culturally defined norm in engineering community, that is, an association of engineers with an ideology of maleness and hard-core technology. From a perspective of 'difference-sameness' (Kvande, 1999), I argue that conflicts arise from the cultural understanding of what count as real engineering.

The conflicts in the identity management process at EE can be related to the understanding of the general conceptions of masculinity and femininity. Students' perceptions on gender relations reflect the feminist description (Harding, 1986; Connell, 1987; Gherardi, 1995) of symbolism of masculinity and femininity in many western cultures, which identifies a close relation between engineering identity and the 'hegemonic masculinity' (1987). This goes in turn with the two kinds of masculinity in engineering as Wajcman (1991) outlines (the more intellectual-focused and the more physical-focused work). Though there are differences between them, the common feature is the control and mastery over technology. In concern with technology, when masculinity is considered as competence (Wajcman, 1991), femininity as an opposition to masculinity (Harding, 1986), should be less technological competent than men. Therefore, in the male-centered community like EE, where practices are established and valued based on male norm, women are considered as different from or the same with men, but not the other way around. Once dealing with machines and technology in a career-oriented profession, female engineering students in hard-core engineering programs are not regarded as feminine any more according to the traditional social ideology of women. They are regarded as different from women in other specializations both by themselves and by their male peers.

According to Salminen-Karlsson (1999), in the current Scandinavian culture, young women are expected to know technology to some extent; however, to work on technology means something different. This study observed the similar situation in the Danish culture, where an increased interest in technology is shown by young women. However, there is still a distance between women and the profession of traditional engineering, which is strongly associated with hardcore technology.

At A&D, conflicts in identity management can be attributed to the way of defining engineering as hardcore technology-related work with representatives of hard-core branches like electronics, construction, and mechanics. As a profession-oriented discipline, engineering education represents the engineering culture, where students are provided with engineering expertise by being trained to think and work in particular ways, which are associated with an image of 'effective', 'rational', and 'well-disciplined engineers' (Hacker, 1989; Berner and Mellstrom, 1997). Instead of adapting to the established principles, new engineering programs like A&D brings in new values like creativity and social concern into engineering competences, which confronts the dominating values in the communities of practice in engineering education. In this way, the entry of A&D into engineering program can be seen as a challenger and violator of the engineering norm.

To conclude, the different pictures regarding gendered learning experiences in the two research sites illustrate how learning cultures in the communities of practice are constructed differently under the influence of gender relations, which in turn, shape individual learning processes. Therefore, the conflicts of identity management in the learning processes in both EE and A&D are not only something to do with individual behavior (as many students claims that 'maybe it is just me who is different'), but should be attributed to the overall conception of what counts as engineering, which is closely linked with hard-core technology and masculine value.

Summary

In this chapter I have discussed the research findings through analysis at three levels: summary of results from the descriptive reports, comparative analysis in relation to research questions, and theoretical analysis. These three levels of analysis provide different angles to look at the findings. The discussion from different angles also leads itself a way to the answers to the research questions, as summarized in the following.

How do engineering students experience their learning processes in a PBL environment?

Relating the findings on the students' experiences in the learning process of studying engineering to the learning principles of PBL Aalborg Model and to the model of understanding learning at three levels (developed in Chapter 3), this study reach the following conclusion. PBL environment plays a positive role in the learning process of engineering students at AAU with respect to: 1) perceiving learning as a process of both taking in and creating knowledge in certain social contexts; 2) valuing the self-direction, active participation, communicative interaction, as well as the reflection on the experiences in the learning activities; 3) developing responsibilities and strategies to manage their own learning; 4) having awareness of self-measurement of learning and appreciate the meaningfulness in the learning process, and 5) preparing themselves with engineering competences for the future work.

In addition, this study also concludes that learning involves a process of identity development. Studying engineering in the PBL environment, which is imitated work place environment, students not only master scientific technical knowledge, but also get opportunities of experiencing actual work practice. In this way, study engineering in the PBL environment, they do not only learn to become an engineering student, but also develop engineering competences and obtain a sense of belonging to the engineering profession. Relating to what Masschelein and Ricken (2003) suggest about the productive feature of power from the perspective of Bildung, I argue that this process of seeking meaningfulness, developing identity and achieving self-transformation is also a process whereby individuals are empowered through learning. The example of A&D also indicates a tendency of change for an engineering community. The professional engineering identity is being challenged by the establishment of new programs like A&D, in which new engineering competences are brought about with the introduction of new contents and the participation of new members.

What are the influences of prevailing gender relations on the learning processes of engineering students in PBL?

Through relating the empirical findings to the analytical framework on gender and learning, this study identifies the influence of gender relations on the organizational culture, which in turn, shapes individual behaviours and interpersonal interactions in the learning processes through participating in the practices of engineering communities. Drawing on the research findings, this study argues that despite the overall gender equality in legislation, the cultural gender contracts still to some extent defines gendered labor division in the society, under the influence of which, communities of practice are doing gender differently. Learning in communities of practice, through participating in different activities, people learn the local languages, practices, values and norms through the social interaction and develop certain identities. From a feminist perspective, this participation is also 'guided' (Rogoff, 1995) by the prevailing gender relations and the gendered values in the community. They need to learn to do gender appropriately through the interaction at interpersonal and individual-context relations, because the communities of practice are doing gender.

Therefore, the culturally defined norm in the engineering community, that is, an association of engineers with an ideology of maleness and hard-core technology, play a role in the learning processes of participants, which leads to the conflicts of managing different identities for some of the members. On the other hand, individuals not only live up to the normative gender relations based on the gender of community (for example, the male norm in the gender culture of EE), but also participate into the engendering of the community culture (as shown in the program of A&D). The diversity in strategies development exemplifies the reciprocity in the individual/community relationship.

By understanding learning as a process of identity development, management and change in a situated community, I see a mutual constructing relation between the learners and the community in the learning process. On one hand, learners conform to the prevailing beliefs, attitudes, norms in the communities (the case of EE), and on the other hand, learners also participate into the establishment and the (re)construction of these beliefs, attitudes and norms (the case of A&D).

10

Conclusion: Is PBL a Recipe to Gender Diversity?

At the starting point of this study, I assumed that the problem-based and project-organized learning environment (PBL), as an example of a student-centered learning context, might lead to different individual experiences than those in traditional lecture-based learning context, as reported in different feminist literature (McIlwee and Robinson, 1992; Hacker, 1989; Berner and Mellstrom, 1997). The positive influences have been confirmed through the report in the previous sections. However, two questions that have been kept in my mind during the research process have not been directly answered: is PBL a friendly environment for engineering education? Is PBL a gender friendly environment?

The introduction of the PBL concept to the curriculum construction in the engineering programs of Aalborg University can be seen as a positive example by locating learning as the core of education and making efforts to provide suitable teaching methods to meet the need of learners. Research findings have shown that a PBL environment provides a supportive milieu in which students of both genders appreciate the meaningfulness of learning, and especially for women to learn technology. Therefore, the PBL Aalborg Model can be regarded as a learning environment that is friendly to students of both genders.

However, the research process also identified that it is more complicated to examine gender issues in engineering education than just defining a friendly learning environment. In general, what can be summarized from this study

Gender and Diversity in a Problem and Project Based Learning Environment, 295–297.

is that a PBL learning environment along with the establishment of new engineering programs with contextualized contents would lead to increased recruitment of women and a substantial level of appreciation of learning. New learning contents in engineering education that have a clear focus on the context and application of technology and close relation to society can be to a great extent attractive for not only female students but also for more male students. Project work in teams and collaborative ways of learning make female students highly motivated and feel that the technical part is less difficult to handle compared with individual studies. Women's special values become visible through the diversity of skills gained from PBL environment such as management, planning, communication and collaboration. This plays a positive role in the increase of learning motivation and self-satisfaction in general, and/or women in particular. However, different pictures from two research cases in this study show the learning culture in the micro-setting is complex than uniformed so that it is a long journey to go before reaching a diverse culture as described by Lewis et al. (2000) where 'everyone feels included, appreciated and valued' and where 'everyone can achieve their potential'.

This research work also suggests that recruitment in terms of increasing numbers is not enough for the improvement of diversity. A solution for increasing the general recruitment of engineering students and for producing better engineers as well as increasing the diversity of the profession calls for a change in engineering education and the profession itself. As addressed by both Jarvis (1992, 2009) and Illeris (2007, 2009) that learning leads to change of both individual and society: learning is both at the heart of all social conformity and also at the heart of all social changes (Jarvis, 1992: 24). Learning leads to ongoing changes in different aspects: changes of individuals as the outcome of learning; changes of engineering competences by broadening the contents; changes of the concepts of engineering and engineers by increasing diversity; changes of public ideology of engineering community and engineers. The emergence of 'the new' is one of the central questions in the philosophy of Bildung as well. These ongoing changes will unavoidably bring impact on the knowledge creation and dissemination process in engineering education. Individuals do not only take in knowledge and develop expected competences, but also participate into the process of creating new knowledge. This changing process can be seen as an empowering process of both individuals and the social structures. As Koller (2003) points out, knowledge is legitimated by being new or by being brought forth through the violation of existing rules.

To conclude, a PBL environment is supportive for the learning process of both male and female engineering students. However, a PBL environment itself is not enough to be used as a recipe for recruiting women to engineering studies. Improving gender diversity in engineering education is more complex than just getting women enrolled in engineering programs; it also involves how to break down the glass wall of engineering profession in general and making engineering education inclusive for all potential groups of students. To achieve that, it requires the appreciation of learning as well as getting prepared for professional practice with diverse new engineering skills; all aspects that include a supportive learning environment, effective teaching and learning methods that will facilitate the development of diverse engineering competencies; and also contextual learning that will relate learning as an engineering student to practicing as an engineer. The improved diversity (for example, with increased female participation in engineering education), in turn, will enrich engineering knowledge, practice, innovative products, values and engineering culture.

References

ABET (Accreditation Board for Engineering and Technology) (2007). Criteria for accrediting engineering programs: Effective for evaluations during the 2008–2008 acreditation cycle. Baltimore, MD: ABET.

Always, J. (1995). The trouble with gender: Tales of the still-missing feminist revolution in sociological theory. *Sociological Theory* 3: 209–228.

Andersen, A. and J. Hansen. (2001). Engineers of tomorrow. In: *New Engineering Competencies — Changing the Paradigm*! Proceedings of the SEFI Annual Conference, Copenhagen.

Anderson, L. and K. Gilbride. (2007). The future of engineering: A study of the gender bias. *Mcgill Journal of Education* 42(1): 103–118.

Antonini, C. (2002). How should we train the next generation of engineers? In: *The Renaissance Engineer of Tomorrow*. Proceedings of 30th SEFI Annual Conference. Firenze, Italy.

Barnett, R. (1994). *The Limits of Competence — Knowledge, Higher Education and Society*. SRHE and Open University Press, Buckingham.

Barrow, H. (1985). *How to Design a Problem-based Curriculum for the Preclinical Years*. Springer, New York.

Barrow, H. (2000). Foreword. In: *Problem-based Learning — A Research Perspective on Learning Interactions*, Evensen, D. and C. Hmelo, (eds.), pp. 1–18. Lawrence Erlbaum Associates Publications, London.

Barsony, J. (2002). Engineer and society. In: *The Renaissance Engineer of Tomorrow*. Proceedings of 30th SEFI Annual Conference. Firenze, Italy.

Bauer, W. (2003). On the relevance of bildung for democracy. *Educational philosophy and Theory* 35(2): 211–226.

Beder, S. (1998). A bit of the rain man in every engineer? *Engineers Australia*, April 1998. p. 57.

Belenky, M., B. Clinchy, N. Goldberger, and J. Tarule. (1986). *Women's Ways of Knowing — The Development of Self, Voice, and Mind*, Basic Books, New York.

Berner, B. (1997). Doing feminist research on technology and society. In: *Gendered Practices: Feminist Studies of Technology and Society*, Berner, B., (eds.), pp. 9–18. Almqvist and Wiksell International, Stockholm.

Berner, B. and U. Mellstrom. (1997). Looking for mister engineer: Understanding masculinity and technology at tow Fin de Siecles. In: *Gendered Practices: Feminist Studies of*

Gender and Diversity in a Problem and Project Based Learning Environment, 299–309.

Technology and Society. Berner, B., (eds.), pp. 39–68. Almqvist and Wiksell International, Stockholm.

Bigge, M. L. and S. S. Shermis. (1999). *Learning Theories for Teachers* (6th edn.). Longman, London.

Blaikie, N. (2000). *Designing Social Research — The Logic of Anticipation.* Polity Press, Cambridge.

Blaikie, N. (1993). *Approaches to Social Enquiry.* Polity Press, Cambridge.

Bowden, J. and F. Marton. (1998). *The University of Learning — Beyond Quality and Competence in Higher Education.* Kogan Page, London.

Brandell, G. (1996). *Gender in Engineering Education.* Centre for Women's Studies, University of Lulea, Lulea.

Brandell, G. et al. (1998). *Encouraging More Women into Computer Science: Initiating a Single-sex Intervention Program in Sweden.* University of Lulea, Lulea.

Brookfield, S. D. (1983). *Adult Learning, Adult Education and the Community.* Milton Keynes, Open University Press.

Brown, J. S., A. Collins, and P. Duguid. (1989). Situated Cognition and the Culture of Learning. *Educational Researcher* 18(1): 32–42.

Christensen, H. P. (2002). Study strategies for engineering students at DTU. In: *The Renaissance Engineer of Tomorrow.* Proceedings of 30th SEFI Annual Conference. Firenze, Italy.

Christensen, H. P. (2004). Creating a learning environment for engineering education. In: *Faculty Development in Nordic Engineering Education,* Kolmos, A. et al., (eds.), pp. 49–66. Aalborg University Press, Aalborg.

Casti, J. L. (1994). *Complexification: Explaining a Paradoxical World through the Science of Surprise.* Harper Collins, New York.

Cockburn, C. (1985). The material of male power. In: *The Social Shaping of Technology.* Donald, M. and J. Wajcman, (eds.), pp. 125–146. Open University Press, Milton Keynes.

Crossley, M. and G. Vulliamy. (1997). *Qualitative Educational Research in Developing Countries: Current Perspectives.* Garland, London.

Curriculum for Women and Technology (CuWaT) Project (1998). *Changing the curriculum — Changing the Balance?* CuWaT Project, Oslo.

Dahms, M. (1997). Female engineering students — a potential in the reform of engineering education? Presentation in: *The International Conference on Engineering for Sustainable Development (ICESD),* Salaa, Tanzania.

Dahms, M. (1998). Transforming the engineering curricula — gender equality in engineering in Denmark. Presentation in *WEPAN National Conference*, Seattle, U.S.A.

Dahms, M. (1999). Gender equity in engineering in Denmark: Still a long way to go. *Journal of Women and Minorities in Science and Engineering* 5(4): 303–309.

Dececchi, T. et al. (1998). A study of barriers to women's engineering education. *Journal of Gender Studies* 7(1).

Denscombe, M. (1998). *The Good Research Guide.* Open University Press, Buckingham.

Dewey, J. (1916). *Democracy and Education.* The Macmillan Company. http://www.ilt.columbia.edu/publications/dewey.html Retrieved 09-10-05.

Dewey, J. (1938). *Experience and Education.* Collier and Kappa Delta Phi, New York.

Dresling, A. (2001). Teaching soft qualifications through a problem based learning curricula. In: *New Engineering Competencies — Changing the Paradigm*! Proceedings of the SEFI Annual Conference, Copenhagen.

Dryburgh, H. (1999). Work hard, play hard. Women and professionalization in engineering — adapting to the culture. *Gender and Society* 13(5): 664–682.

Deuk, J. E. (2000). Whose group is it, anyway? Equity of student discourse in problem-based learning (PBL). In: *Problem-based learning — A Research Perspective on Learning Interactions*, Evensen, D. and C. Hmelo, (eds.), pp. 75–108. Lawrence Erlbaum Associates Publications, London.

Du, X. Y. (2001). *Swans or Ugly Ducklings?Lives of Female Engineering Students in a Male-centred Technological World.* Linköping University, Linköping.

Du, X. Y. (2003). Problem-based learning — gender friendly? In: *The Proceedings of the 11th International Conference on Gender And Science And Technology (GASAT 11)*, Mauritius.

Du, X. Y. (2005). Gendered features of learning and identity development. In: *Projederfaringer-fra 'Get a Life, Engineer!'* The Danish Engineers Society, Copenhagen.

Du, X. Y. (2006a). Bringing new values in engineering education — gendered and learning in PBL. Thesis [PhD]. Aalborg University.

Du, X. Y. (2006b). Gendered practices of constructing an engineering identity in a problem-based learning environment. *European Journal of Engineering Education* 31(1): 35–44.

Du, X. Y. (2006c). Bildung and identity construction in engineering education. In: *Engineering Science, Skills and Bildung*, Christensen, J., L. B. Henriksen, and A. Kolmos, (eds.), pp. 147–164, Aalborg University Press.

Du, X. Y. and A. Kolmos. (2007). Gender inclusiveness in engineering education — is problem based learning environmenta recipe? *Conference Proceeding for American Society for Engineering Education (ASEE) Annual Conference*, June 24–27, Honolulu, USA.

Du, X. Y., E. de Graaff, and A. Kolmos. (2009). PBL — diversity in research questions and methodologies. In: *Research on PBL Practice in Engineering Education.* Du, X. Y., E. de Graaff, and A. Kolmos, (eds.), pp. 1–7, Rotterdam: Sense Publishers.

Du, X. Y. and A. Kolmos. (2009). Increasing the diversity of engineering education — a gender analysis in a PBL context. *European Journal of Engineering Education* 34(5): 425–437.

Edelman, B. (1997). The girls are no ordinary girls! In: *Gendered Practices: Feminist Studies of Technology and Society.* Berner, B. (ed.), pp. 19–38. Almqvist and Wiksell International, Stockholm.

Eisner, E. W. (1991). *The Enlightened Eye: Qualitative Inquiry and the Enhancement of Educational Practice.* New York, NY: Macmillan.

Engineering Council (2000). *The Universe of Engineering — A U.K. Perspective.* A report prepared by a joint Royal Academy of Engineering/Engineering Council Working Group, Published by Engineering Council, London.

Engineers Australia. (2005). *Engineers Australia Policy on Accreditation of Professional Engineering Programs.* Canberra: Engineers Australia.

ENAEE (European Network for Accreditation of Engineering Education) (2005). EUR-ACE framework standards for the accreditation of engineering programmes. Brussels: European Federation of National Engineering Associations.

Felder, R. M. (2006). A whole new mind for a flat world. *Chemical Engineering Education* 40(2): 96–97.

Fink, F. (2001). Integration of Work Based Learning in Engineering Education. In: *Proceedings of 31st ASEE/IEEE Frontiers in Education Conference (FIE01)*, Reno, Nevada.

Fish, M. D. (1995). Changing the culture of engineering at cornell: Slogan foundation initiatives. In: *The proceedings of 1995 WEAPAN National Conference*, pp. 223–226.

Flannery, D. and E. Hayes. (2000). Women's learning: A kaleidoscope. In: *Women as Learners — the Significance of Gender in Adult Learning*. Hayes, E. and F. Daniele, (eds.) Jossey-Bass Publishers, San Francisco.

Flick, U. (2002). *An Introduction to Qualitative Research* (2nd edn.). SAGE Publication, London.

Fox, M. F., G. Sonnert, and I. Nikiforova. (2009). Successful programs for undergraduate women in science and engineering: Adapting versus adopting the institutional environment. *Res High Educ* 50: 333–353.

Franchetti, M., T. Ravn, and V. Kuntz. (2010). Retention and recruitment programs for female undergraduate students in engineering at the university of Toledo, Ohio, USA. *Journal of STEM Education* 11(5 and 6): 25–31.

Frize, M. (1993). Reflections on the engineering profession: Is it becoming friendlier for women? *CSME Bulletin* (12–14) June, 12–14.

Fromentin, A. and D. Werra. (2001). Emerging competencies in engineering. In: *New Engineering Competencies — Changing the Paradigm!* Proceedings of the SEFI Annual Conference, Copenhagen.

Gephart, R. (1999). Paradigms and research methods. *Research Methods Forum,* Vol. 4. Academy of Management, Research Methods Division.

Gherardi, S. (1995). *Gender, Symbolism and Organizational Culture.* SAGE Publications, London.

Giddens, A. (2001). *Sociology*. Polity Press, Cambridge.

Gilligan, C. (1982). *In a Different Voice: Psychological Theory and Women's Development.* Harvard University Press, Cambridge, Mass.

Goldberger, N. (1996). Looking backward, looking forward. In: *Knowledge, Difference, and Power — Essays Inspired by Women's Ways of Knowing,* Goldberger, N., J. Tarule, B. Clinchy, and M. Belenky, (eds.), pp. 1–24. Basic Books, New York.

Gordon, T. et al. (2001). *Ethnographic Research in Educational Settings*, In: Atkinson, Paul et al., (eds.), *Handbook of Ethnography*, Sage Publications, London.

Graaff, E. de (1994). Problem-based learning in engineering education. In: *Project-organized Curricula in Engineering Education*. SEFI cahier No. 4, Brussels.

Graaff, E. de and R. Cowdroy. (1997). Theory and practice of educational innovation: Introduction of problem based learning in architecture. *International Journal of Engineering Education* 13(3).

Graaff, E. de et al. (2001). Research as learning paradigm. In: *New Engineering Competencies — Changing the Paradigm*! Proceedings of the SEFI Annual Conference, Copenhagen.

Graaff, E. de, and A. Kolmos. (2003). Characteristics of problem-based learning, *International Journal of Engineering Education* 19(5): 657–662.

Guba, E. G. and Y. S. Lincoln. (1994). Competing paradigms in qualitative research. In: *Handbook of Qualitative Research*, in Denzin, N. K. and Y. S. Lincoln, (eds.), pp. 105–117, Newbury Park, CA.

Habermas, J. (1984). *The Theory of Communicative Action Vol. 1–Reasons and the Rationalition of Society.* Boston.

Hacker, S. (1990). *Doing It the Hard Way — Investigations of gender and technology.* Unwin Hyman, Boston.

Hacker, S. (1989). *Pleasure, Power, and Technology: Some Tales of Gender, Engineering and the Cooperative Workplace.* Unwin Hyman, Boston.

Hadi, M. N. S. (2001). Challenges facing engineering. In: *New Engineering Competencies — Changing the Paradigm*! Proceedings of the SEFI Annual Conference, Copenhagen.

Harding, S. (1986). *The Science Question in Feminism.* Open University Press, Milton Keynes.

Harding, S. (1987). *Feminism and Methodology.* Bloomington: Indiana University Press.

Harding, S. (1991). *Whose Science? Whose Knowledge? Thinking from Women's Lives.* Open University Press, Milton Keynes.

Harding, S. (1996). Gendered ways of knowing and the epistemological crisis of the west. In: *Knowledge, Difference, and Power — Essays Inspired by Women's Ways of Knowing,* Goldberger, N., J. Tarule, B. Clinchy, and M. Belenky, (eds.), pp. 431–454. Basic Books, New York.

Hayes, E. (2000). Social contexts. In: *Women as Learners — The Significance of Gender in Adult Learning*, Hayes, E. and Daniele, (eds.), pp. 23–52, Jossey-Bass Publishers, San Francisco.

Henwood, F. (1996). WISE choices? Understanding occupational decision-making in a climate of equal opportunities for women in science and technology. *Gender and Education* 8(2): 199–214.

Henwood, F. (1998). Engineering difference: Discourses on gender, sexuality, and work in a college of technology. *Gender and Education* 10(1): 35–49.

Henriksen, L. B. (2006). Engineers and bildung. In: *Engineering Science, Skills and Bildung*, Kolmos et al. (eds.), Aalborg University Press, Aalborg.

Hermanussen, R. and C. Booy. (2002). Equal opportunity in higher technical education: Past, present and future. *International Journal of Engineering Education* 18(4): 452–457.

Hernaut, K. (2002). ICT Curricula for the renaissance engineer in the 21st century. In: *The Renaissance Engineer of Tomorrow.* Proceedings of 30th SEFI Annual Conference. Firenze, Italy.

Hirdman, Y. (1990). The gender system: Theoretical reflections on the social subordination of women. *The Study of Power and Democracy in Sweden, English Series, Report No. 40.* Maktutredningen, Uppsala.

Hmelo, C. and D. Evensen. (2000). Introduction. In: *Problem-based Learning — a Research Perspective on Learning Interactions*, Evensen, D. and C. Hmelo, (eds.), pp. 1–18. Lawrence Erlbaum Associates Publications, London.

Hmelo, C. and X. Lin. (2000). Becoming self-directed learners: Strategy development in problem-based learning. In: *Problem-based Learning — a Research Perspective on Learning Interactions*, Evensen, D. and C. Hmelo, (eds.), pp. 227–250. Lawrence Erlbaum Associates Publications, London.

Hodgkinson, L. and L. Hamill. (2006). Engineering careers in the UK: Still not what women want? *Industry and Higher Education*, Dec 403–412.

Hoepfl, M. C. (1997). Choosing qualitative research: A primer for technology education researchers, *Journal of Technology Education* 9(1).

Hohle, J. (2004). The AAU model seen from a foreign professors' point of view. In: *The Aalborg PBL Model — Progress, Diversity and Challenges*, Kolmos, A., F. Fink, and L. Krogh, (eds.), pp. 381–390. Aalborg University Press, Aalborg.

Huitt, W. and J. Humel, (2003). Piaget's theory of cognitive development. *Educational Psychology Interactive.* Valdosta, GA: Valdosta State University.

IDA (2002). *FremtidensIngeniorprofiler*. Publication of the Engineering Society in Denmark (IDA).

Ihsen, S. and S. Gebauer. (2009). Diversity issues in the engineering curriculum. *European Journal of Engineering Education* 34(5): 419–424.

Illeris, K. (2009). A comprehensive understanding of human learning. In: *Contemparary Theories of Learning*. Illeris, K. (ed.). London: Routledge. 7–20.

Jarvis, P. (1987). *Adult Learning in the Social Context.* Croom, Helm, London.

Jarvis, P. (1992). *Paradoxes of Learning — On Becoming an Individual in Society.* Jossey-Bass, San Francisco.

Jarvis, P. (1995). *Adult and Continuing Education: Theory and Practice* (2nd edn.). Routledge, London.

Jarvis, P., J. Holfore, and C. Griffin. (1998). *The Theory and Practice of Learning.* Kogan Page Limited, London.

Jarvis, P. (2001). *Universities and Corporate Universities.* Kogan Page, London.

Jarvis, P. (2003). *Adult and Continuing Education — Major Themes in Education.* Routledge, London.

Jarvis, P. (2009). Learning to be a person in society: Learning to be me. In: *Contemparary Theories of Learning*. Illeris, K. (ed.). London: Routledge. 21–34.

Jensen, A. A. and H. Baekkelund. (2004). Back to the future — theory and practice in adult practitioners' problem oriented project work. In: *The Aalborg PBL Model — Progress, Diversity and Challenges,* Kolmos, A., F. Fink, and L. Krogh, (eds), pp. 283–300. Aalborg University Press, Aalborg.

Kim, K. A., A. J. Fann, and K. O. Misa-Escalante. (2011). Engaging women in computer science and engineering: Promising practices for promoting gender equity in undergraduate research experiences. *ACM Transactions on Computing Education* 11(2): 8:1–8:19.

Kjaersdam, F. (1990). Problem-oriented higher education for engineering and technology. In: *International Symposium on Higher Engineering Education.* Oxford: Pergaman Press.

Kjaersdam, F. (1993). Evaluation of project-organized engineering education. *European Journal of Engineering Education* 18(4).

Kjaersdam, F. (1994). The Aalborg experiment — tomorrow's engineering education. *European Journal of Engineering Education* 19(3).

Kjaersdam, F. and S. Enemark. (1994). *The Aalborg Experiment — Project Innovation in University Education.* Aalborg University Press, Aalborg.

Kjaersdam, F. (2002). The problem solving renaissance engineer. In: *The Renaissance Engineer of Tomorrow.* Proceedings of 30th SEFI Annual Conference. Firenze, Italy.

Kofoed, L. et al. (2004). Teaching process competencies in a PBL curriculum. In: *The Aalborg PBL Model — Progress, Diversity and Challenges,* Kolmos, A., F. Fink, and L. Krogh, (eds.), pp. 331–348. Aalborg University Press, Aalborg.

Kolb, D. (1984). *Experiential Learning — Experience as the Source of Learning and Development.* Prentice Hall PER, New Jersey.

Koller, H. C. (2003). Bildung and radical plurality: Towards a redefinition of bildung with reference to J. F. Lyotard. *Educational Philosophy and Theory* 35(2): 155–166.

Kolmos, A. (1991). *Open Learning in Engineering Education — A Future Trend?* Contributions — GASAT 6 International Conference, Australia.

Kolmos, A. (1992). *Metacognitive Aspects in A Group-based Project Work at Technical Universities.* Contributions GASAT, The Netherlands.

Kolmos, A. (1996). Reflection on project work and problem-based learning. *European Journal of Engineering Education* 21(2).

Kolmos, A. (1999). Progression of collaborative skills. In: *Themes and Variations in PBL*, Conway, J. and A. Williams, (eds.), vol. 1, pp. 129–138. Callaghan, NSW: Australian Problem Based Learning Network.

Kolmos, A. et al. (2001). Organization of staff development — strategics and experiences. *European Journal of Engineering Education* 26(4): 329–342.

Kolmos, A. (2002). Facilitating change to a problem-based model, *The International Journal for Academic Development* 7(1): Routhledge.

Kolmos, A. (2002a). Future competencies and learning methods in engineering education. In: *The Proceedings of the 6th Baltic Region Seminar on Engineering Education*, Wismar/Warnemunde, Germany.

Kolmos, A., F. Fink, and L. Krogh. (2004). The Aalborg model — problem-based and project-organized learning. In: *The Aalborg PBL Model — Progress, Diversity and Challenges*, Kolmos, A., F. Fink, and L. Krogh, (eds.), pp. 9–18. Aalborg University Press, Aalborg.

Kolmos, A. and O. Vinther. (2004). Faculty development strategies at the danish engineering education. In: *Faculty Development in Nordic Engineering Education*, Kolmos, A. et al., (eds.), pp. 123–132. Aalborg University Press, Aalborg.

Kolmos, A. et al. (2004). Perspectives on nordic faculty development. In: *Faculty Development in Nordic Engineering Education*, Kolmos, A. et al., (eds.), pp. 5–12. Aalborg University Press, Aalborg.

Kolmos, A. (2006). Engineering knowledge skills and identity. In: *Engineering Science, Skills and Bildung*, Kolmos et al., (eds.), Aalborg University Press, Aalborg.

Koschmann, T. et al. (2000). When is a problem-based tutorial not a tutorial? Analyzing the tutor's role in the emergence of a learning issue. In: *Problem-based Learning — A Research Perspective on Learning Interactions*, Evensen, D. and C. Hmelo, (eds.), pp. 53–74. Lawrence Erlbaum Associates Publications, London.

Krueger, R. A. (1994). *Focus Groups: A Practical Guide for Applied Research.* SAGE Publication, London.

Kvale, S. (1996). *Interviews — An Introduction to Qualitative Research Interviewing.* SAGE Publication, London.

Kvande, E. (1999). In the belly of the beast — constructing femininities in engineering organizations. *The European Journal of Women's Studies* 6: 305–328.

Kvande, E. and B. Rasmussen. (1994). Men in male-dominated organizaitons and their encounter with women intruders. *Scandinavian Journal of Management* 10(2): 164–175.

Lather, P. (1991). *Feminist Research in Education: Within/Against.* Deakin University Press, Geelong, Victoria.

Lave, J. (1988). *Cognition in Practice: Mind, Mathematics, and Culture in Everyday Life.* Cambridge University Press, Cambridge.

Lave, J. and E. Wenger. (1991). *Situated Learning — Legitimate Peripheral Participation.* Cambridge University Press, Cambridge.

Lave, J. (1997). Learning, apprenticeship, social practice. *Journal of Nordic Educational Research* 17.

Lave, J. (2009). The practice of learning. In: *Contemparary Theories of Learning*. Illeris, K. (ed.), pp. 200–208. London: Routledge.

Lewis, S. (1993). Including gender in higher education science and engineering courses. In: *The Proceedings of GASAT 7 International Conference*, pp. 662–669, Canada.

Lincoln, Y. and E. Guba. (1985). *Naturalistic Inquiry.* Sage Publications, Beverly Hills, CA.

Litosseliti, L. (2003). *Using Focus Groups in Research.* Continuum, London.

Lucena, J. and G. Downey. (2001). Engineering cultures: Addressing challenges of globalization to engineering education through humanand cultural-centred problem-solving. In: *New Engineering Competencies — Changing the Paradigm*! Proceedings of the SEFI Annual Conference, Copenhagen.

Maher, F. (1996). Women's ways of knowing in women's studies, feminist pedagogies, and feminist theory. In: *Knowledge, Difference, and Power — Essays Inspired by Women's Ways of Knowing,* Goldberger, N., J. Tarule, B. Clinchy, and M. Belenky, (eds.), pp. 148–174. Basic Books, New York.

Male, S. A., M. B. Bush, and K. Murray. (2009). Think engineer, think male? *European Journal of Engineering Education* 34(5): 455–464.

Marshall, C. (1997). Dismantling and reconstructing policy analysis. In: *Feminist Critical Policy Analysis: A Perspective from Primary and Secondary Schooling*, Marshall, C. (eds.), (vol. 1, pp. 1–34), Falmer, London.

Marshall, C. and G. B. Rossman. (1999). *Designing Qualitative Research* (3rd ed). SAGE, London.

Masschelein, J. and N. Richen. (2003). Do we (still) need the concept of bildung? *Educational Philosophy and Theory* 35(2): 139–144.

McIlwee, S. and G. Robinson, (1992). *Women in Engineering: Gender, Power, and Workplace Culture.* State University of New York Press, Albany.

Mellstrom, U. (1995). *Engineering Lives: Technology, Time and Space in a Male-centred World.* Linköping University, Linköping.

Miles, M. B. and A. M. Huberman. (1993). *Qualitative Data Analysis: A Sourcebook for New Methods.* SAGE, Newbury Park, CA.

Miliszewska, I. and A. Moore. (2010). Encouraging girls to consider a career in ICT: A review of strategies. *Journal of Information Technology Education: Innovations in Practice* 9: 143–166.

Mishler, E. G. (1986). *Research Interviewing — Context and Narrative.* Harvard University Press, Cambridge.

Mosby, E. (2005). Curriculum development for project-oriented and problem-based learning (POPBL) with emphasis on personal skills and abilities. *Global Journal of Engineering Education* 9(2): 121–128.

National Academy of Engineering (NAE). 2004. *The Engineer of 2020: Visions of Engineering in the New Century*. Washington, DC: National Academies Press.

Patton, M. Q. (1990). *Qualitative Evaluation and Research Methods* (2nd edn.) SAGE Publications, Newbury Park, CA.

Phipps, A. (2002). Engineering women: The 'Gendering' of professional identities. *International Journal of Engineering Education* 18(4): 409–414.

Phillips, D. C. and J. F. Soltis. (1998). *Perspectives on Learning*. Teachers College Press, New York.

Powell, A., B. Bagilhole, and A. Dainty. (2009). How women engineers do and undo gender: Consequences for gender equality. *Gender, Work and Organization* 16(4): 411–428.

Punch, K. F. (1998). *Introduction to Social Research.* Sage Publications, London.

Reason, P. (1988). *Human Inquiry in Action*. Sage Publications, London.

Reeves, T. (1996). *Educational Paradigms.* http://www.itech1.coe.uga.edu/reeves.html Retrieved July 22nd, 2004.

Riley, D., A. L. Pawley, J. Tucker, and G. D. Catalano. (2009). Feminisms in engineering education: Transformative possibilities. *NWSA Journal* 21(2): 21–40.

Roberts, P. and M. Ayre. (2002). Did she jump or was she pushed? A study of women's retention in the engineering workforce. *International Journal of Engineering Education* 18(4): 415–421.

Rogers, A. (2002). *Teaching Adults* (3rd edn.). Open University Press, Philadelphia.

Rogoff, B. (1994). Developing understanding of the idea of communities of learners. *Mind, Culture and Activity* 1(4): 209–229.

Rogoff, B. (1995). Observing sociocultural activity on three planes: Participatory appropriation, guided participation, and apprenticeship. In: *Sociocultural Studies of the Mind.* Wertsch, J. V., P. Del Rio, and A. Alvarez, (eds.), pp. 139–164. Cambridge University Press, Cambridge.

Rogoff, B. (1997). Evaluating development in the process of participation: Theory, methods, and practice building on each other. In: *Change and Development: Issues of Theory, Method, and Application.* Amsel, E. and Renninger (eds.), pp. 265–285. Lawrence Erlbaum Associates, Mahwah, NJ.

Rosser, S. (1996). Forstering the advancement of women in the sciences, mathematics and engineering. In: *The Equity Equation.* Davis, C.-S. et al. (eds.), Jossey-Bass, San Francisco.

Rossman. G. B. and S. F. Rallis. (1998). *Learning in the Field: And Introduction to Qualitative Research.* SAGE, Thousand Oaks,CA.

Sagebiel, F. and J. Dahmen. (2006). Masculinities in organizational cultures in engineering education in Europe: Results of he European Union project Wom Eng. *European Journal of Engineering Education* 31(1): 5–14.

Salminen-Karlsson, M. (1997). Reforming a masculine bastion: State-supported reform of engineering education. In: *Gendered Practices: Feminist Studies of Technology and Society.* Berner, B. (ed.), pp. 187–204. Almqvist and Wiksell International, Stockholm.

Salminen-Karlsson, M. (1999). *Bringing Women into Computer Engineering: Curriculum Reform Processes at Two Institutes of Technology.* Linköping University, Linköping.

Salminen-Karlsson, M. (2002). Gender inclusive computer engineering education — two attempts at curriculum change. *International Journal of Engineering Education* 18(4): 430–437.

Schaafer, A. (2003). Imaginary horizons of educational theory. *Educational Philosophy and Theory* 35(2): 189–200.

Sheppard, S. D., K. Macatangay, A. Colby, and W. M. Sullivan. (2009). *Educational Engineers, Designing for the Future of the Field.* San Francisco: Jossey-Bass.

Schmidt, H. G. and H. C. Moust. (2000). Factors affecting small-group tutorial learning: A review of research. In: *Problem-based Learning — A Research Perspective on Learning Interactions*, Evensen, D. and C. Hmelo, (eds), pp. 19–52. Lawrence Erlbaum Associates Publications, London.

Schwandt, T. A. (1994). Constructivist, interpretivist approaches to human inquiry. In: *Handbook of Qualitative Research*, Denzin, N. K. and Y. S. Lincoln, (eds.) pp. 118–137. Newbury Park, CA.

Schäfer, A. I., 2006. A new approach to increasing diversity in engineering at the example of women in engineering. *European Journal of Engineering Education* 31(6): 661–671.

Seymour, E. and N. M. Hewitt. (1997). *Talking about Leaving. Why Undergraduates Leave the Sciences.* Westview Press, Boulder.

Shull, J. and M. Weiner. (2002). Thinking inside the box: Self-efficacy of women in engineering, *International Journal of Engineering Education* 18(4): 438–446.

Silverman, D. (2000). *Doing Qualitative Research — A Practical Handbook.* SAGE Publication, London.

Silverman, D. (2001). *Interpreting Qualitative Data — Methods for Analysing Talk, Text and Interaction* (2nd edn.). SAGE Publication, London.

Sim, G. and R. Hensman. (1994). Science and technology: Friends or enemies of women? *Journal of Gender Studies* 3(3): 277–287.

Spradley, J. P. (1979). *Ethnographic Interview.* Holt, Rinehart and Winston, New York.

Spradley, J. P. (1980). *Participant Observation.* Holt, Rinehart and Winston, New York.

Steier, F. (1991). *Research and Reflexivity.* SAGE Publication, London.

Stonyer, H. (2002). Making engineering students — making women: The discursive context of engineering education, *International Journal of Engineering Education* 18(4): 392–399.

Sulaiman, N. F. and H. AlMuftah. (2010). A qatari perspective on women in the engineering pipeline: An exploratory study. *European Journal of Engineering Education* 35(5): 507–517.

Sundin, E. (1997). Gender and technology: Mutually constituting and limiting. In: *Gendered Practices: Feminist Studies of Technology and Society.* Berner, B. (eds.), pp. 249–268. Almqvist and Wiksell International, Stockholm.

Tarule, J. (1996). Voices in dialogue: Collaborative ways of knowing. In: *Knowledge, Difference, and Power — Essays Inspired by Women's Ways of Knowing*. Goldberger, N., J. Tarule, B. Clinchy, and M. Belenky, (eds.), pp. 274–304. Basic Books, New York.

Tennant, M. (1997). *Psychology and Adult Learning* (2nd edn.). Routledge, London.

Thaler, A. and I. Zorn. (2010). Issues of doing gender and doing technology — music as an innovative theme for technology education. *European Journal of Engineering Education* 35(4): 445–454.

Tisdell, E. J. (2000). Feminist pedagogies. In: *Women as Learners — the Significance of Gender in Adult Learning,* Hayes, E. and F. Daniele, (eds.), pp. 155–184. Jossey-Bass Publishers, San Francisco.

Tonso, K. (1996a). Student learning and gender. *Journal of Engineering Education* 85(2): 143–150.

Tonso, K. (1996b). The impact of cultural norms on women. *Journal of Engineering Education* 86(3): 217–225.

Tonso, K. (2007). On the outskirts of Engineering: Learning identity, gender and power via engineering practice. Rotterdam: Sense Publishers.

Van, M. J. (1988). *Tales of the Field.* University of Chicago Press, Chicago.

Varma, R. and H. Hahn. (2008). Gender and the pipeline metaphor in computing. *European Journal of Engineering Education* 33(1): 3–11.

Vernon, D. T. and R. L. Blake. (1993). Does problem-based learning work? A meta-analysis of evaluative research. *Academic Medicine* No. 68, pp. 550–563.

Vidal, R. (2001). Creativity for engineers. In: *New Engineering Competencies — Changing the Paradigm!* Proceedings of the SEFI Annual Conference, Copenhagen.

Wajcman, J. (1991). *Feminism Confront Technology.* Polity Press, Cambridge.

Wajcman, J. (2000). Reflections on gender and technology studies: What state is the art? *Social Studies of Science -SSS and SAGE Publications*, pp. 447–464, London.

Webster, J. (1997). Information technology, women and their work. In: *Gendered Practices: Feminist Studies of Technology and Society.* Berner, B. (ed.), pp. 141–156. Almqvist and Wiksell International, Stockholm.

Wenger, E. (1998). *Communities of Practice — Learning, Meaning, and Identity.* Cambridge University Press, Cambridge.

Wenger, E. (2004). *Learning for a Small Planet — A Research Agenda.* on line (http://www.ewenger.com/ewbooks.html). Retrieved May, 2009.

Wenger, E. (2009). A social theory of learning. In: *Contemparary Theories of Learning.* Illeris, K. (ed.), pp. 209–218. London: Routledge.

Wertsch, J. V. (1985). *Vygotsky and the Social Formation of Mind.* Harvard University Press, Cambridge.

West, C. and H. Zimmerman. (1991). Doing gender. In: *The Social Construction of Gender*, Lorber, J. and A. Farrell, (ed.). SAGE, Newbury Park.

William, A. (2002). Let's TWIST: Creating a conductive learning environment for women. *International Journal of Engineering Education* 18(4): 447–451.

Wimmer, M. (2003). Ruins of bildung in a knowledge society: Commenting on the debate about the future of bildung. *Educational Philosophy and Theory* 35(2): 167–188.

Woolfolk, A. (1987). *Educational Psychology* (3rd edn.). Prentice-Hall, Inc, New Jersey.

Wright, P. (1994). *Introduction to Engineering* (2nd edn.). John Wiley and Sons INC, New York.

Vygotsky, L. S. (1978). *Mind in Society: The Development of Higher Psychological Processes.* Harvard University Press, Cambridge, MA.

Zimmerman and Lebeau (2000). A Commentary on Self-Directed Learning. In: *Problem-based Learning — a Research Perspective on Learning Interactions*, Evensen, D. and C. Hmelo, (eds.), pp. 299–315. Lawrence Erlbaum Associates Publications, London.

Index

Authors' Biography

Xiang-Yun Du, PhD, is a professor at Department of Learning and Philosophy, Aalborg University, Her main research interests and publications focus on innovative teaching and learning in higher education, in particular, Problem Based and Project Based Learning methodology in diverse social, cultural, educational and professional contexts. Within PBL related research field, she has been involved in more than 20 research projects on educational change, curriculum development, pedagogy development, intercultural studies, and gender and diversity research.